# Born in the Jungles of Burma

# Born in the Jungles of Burma: Behind Enemy Lines in the China-Burma-India Theater of Operations

By

Andrew Wax

CAMBRIDGE SCHOLARS
PUBLISHING

Born in the Jungles of Burma:
Behind Enemy Lines in the China-Burma-India Theater of Operations,
by Andrew Wax

This book first published 2010

Cambridge Scholars Publishing

12 Back Chapman Street, Newcastle upon Tyne, NE6 2XX, UK

British Library Cataloguing in Publication Data
A catalogue record for this book is available from the British Library



ISBN (10): 1-4438-2414-3, ISBN (13): 978-1-4438-2414-9

# TABLE OF CONTENTS

# ACKNOWLEDGMENTS

There is no question that I would not have reached this point of completion of this program without the total support and faith and understanding of my wife, Mindi and my son, Justin. Mindi has been solely responsible for my seeing this program through and my son's continual interest and confidence have been the galvanizing factors that allowed me to reach this point in my academic career. Without such uncompromising support, I never would have been able to reach the finish line.

From a purely academic point of view, the individuals who were most responsible for not only recognizing my efforts and capabilities, but always there to spur me on need to be recognized. Foremost in terms of their efforts on my behalf are Drs. Jeffrey Kinkley, Frank Coppa, and Mauricio Borrero, all of the St. John's History Department. Their generosity of time, insight, and compassion made my experience considerably less onerous and tedious than it might otherwise have been. Given that at the beginning of my quest I had more doubts than can be imagined, their respective counseling and guidance enabled me to persevere.

In addition, one individual, Dr. Richard Greenwald, has been my advocate for these past few years and his unique perspective and understanding of my own particular eccentricities were invaluable in persevering through the rough patches of my doctoral program.

# CHAPTER ONE

# ORIGINS

By the 1960s, the United States Army had developed a method of troop deployment that could place forces in any seemingly inaccessible place via aircraft while providing for their every need until their missions were accomplished. Air mobility became an accepted doctrine for the Army, and although the helicopters were vulnerable to ground fire, they still enabled the air cavalry units to be highly effective and versatile for whatever job they were called upon to do. This air mobility development required a beginning. It did not just happen in the jungles of Vietnam. When and where did United States' forces theorize and implement the concept of air mobility? And how did the commanders take theory and put it into actual use? The following pages trace this development, revealing in considerable detail a story that has remained beneath the radar for over six decades.

The introduction of heavier than air craft in appreciable numbers was during the First World War, and this is where the story begins. During World War One, General William Mitchell of the United States Army Air Service devised a plan to end the stalemate on the Western Front. He would drop thousands of troops by parachute behind enemy lines and then use them to attack the rear of the German trench lines. Given the relatively primitive nature of aircraft, their minimal ranges and carrying capacities, and the newness of parachutes available in 1912, this audacious plan had merit, but little practicality. Happily, for the men who may have been earmarked for this adventure, the war came to an end before the plan could be implemented.[1]

After the war, William Mitchell wrote books advocating the development of air power. However, nowhere does he mention the dropping of troops by parachute and the necessary supply and support of these troops. His recommendation to General John Pershing, Commanding General of the American Expeditionary Force, never went beyond the talking stage of putting troops behind enemy lines and providing for them. Could Mitchell perhaps have seen no purpose in pursuing that line of thinking and let the matter drop? Had the war continued into 1919 and beyond, it would then have required a great deal more planning, including the proper support of thousands of troops behind enemy lines or caught behind enemy lines without a direct line of communication or supply.

Prior to 1940, the use of paratroopers and/or glider troops placed behind enemy lines was little thought of until the German Army did it. That development meant that all major armies of the 1940s had to have paratroopers of some kind. After the 1941 campaign against the British on Crete, following a successful use of air-landed glider-borne troops and paratroopers, Adolf Hitler forbade the future use of these troops because of the enormous number of casualties sustained in Crete.[2]

On October 23, 1941, a letter from the Civil Aeronautics Administration to Brigadier General Oliver P. Echols, then Acting Chief of the Air Corps, informed him that following Congressional approval, and "based on the alleged success of the Germans in landing men and supplies in Crete by means of trains of gliders towed by airplanes, that it now became necessary for the Army Air Corps to provide a reservoir of pilots trained in handling gliders having flying characteristics similar to those required of troop carrying gliders."[3]

The United States Army, with a paratrooper school at Fort Bragg in North Carolina and at Wright Field in Dayton, Ohio, did carry out some experiments concerning parachutes and paratroopers in 1940 and 1941. However, did the Army at any time give thought to the possibility of putting troops down somewhere far behind enemy lines? And following that deployment, what efforts would be necessary to sustain those forces so as to prevent them from being annihilated shortly thereafter?

By the middle of 1942, both the British and Americans had laid the groundwork for the development of airborne units: the British, the 1st and 6th Airborne Divisions, and the United States, the 82nd and 101st Airborne Divisions, with more to follow. These units played an important role not only in the invasion of Sicily in July of 1943, but they would also serve significantly in the invasion of Normandy in June of 1944. However, in both campaigns, these airborne troops were not put behind enemy lines for more than a few days at most. Their missions were to cut off reinforcements, capture strategic bridges or crossroads, or disrupt the rear just long enough to enable the ground troops to link up with them.

Even in the ill-conceived plan designed and developed by Field Marshall Bernard Law Montgomery known as Operation Market-Garden to take place in northern Holland in September of 1944, the airborne troops of the American 82nd and 101st and the British 1st Airborne Division were supposed to hold onto their respective bridgeheads only a few days until the arrival of ground forces. These lightly-armed paratroopers could not possibly withstand a heavy armored response on the part of the Germans, or keep them away from the five bridges, the capture of which was the key to success in this operation. As witnessed by the near complete destruction of the 1st British Airborne Division, the Allies showed an inability to secure the wellbeing of the paratroopers behind enemy lines.[4]

Yet, the European Theater of Operations was first in line for most U.S. military resources (following discussions with Winston Churchill while he was at the White House over the Christmas holiday in 1941). Given that the flying and dropping of troops behind enemy lines was an exceedingly difficult task fraught with enormous danger, what chances of success did airborne units have if they were down the line in allocation of scarce resources as was the case for all Allied units in the China-Burma-India Theater of Operations? There was no question that the bulk of Allied resources were focused on the European and Pacific theaters. Was it at all possible for troops placed behind enemy lines to succeed in any other theater of operations, with substantially fewer resources?

To answer that question, it is necessary to return to the United States in the period just prior to the war in Europe and investigate what, if anything, was done by the United States Army in the realm of supply and support of troops behind enemy lines. Was it possible for the men of the United States Army Air Force to be sent to an extremely inhospitable place and commence operations with little or no training, or did such units at least acquire a modicum of understanding of and capability to perform their mission prior to their orders to ship out?

Based on the available records, it would seem that little if anything had ever been done or even conceived of about the supply and support of troops behind enemy lines solely by air transport. However, in the case of developing a new type of aircraft that might prove useful for the purpose of just such a mission, in 1924 the United States Army Air Corps received a proposal for the urgent need of developing helicopters. This proposal pointed out that from a military standpoint, "Helicopters will first of all be the connecting links between different parts of an army. By their ability of being able to fly over rivers and mountains and to land almost anywhere, helicopters will establish communication between military posts that cannot be reached by horse, motor-cycle, automobile or aeroplane.[sic]"[5] What the proposal does not point out, however, is that a slow-moving, low-flying aircraft would most likely be extremely vulnerable to ground fire and certainly could not survive in any but an environment of complete air superiority.

Although this proposal might have seemed farfetched, the experience of the United States Marines in Nicaragua between 1926 and 1933 added impetus to the development of such an aircraft. Marine garrisons had been placed throughout the Nicaraguan countryside ostensibly to protect American personnel and interests. Some of the more isolated outposts in the remote northern province of Nueva Segovia required a different method so as to be kept in contact with their base and be supplied on a regular basis. Marine aviators were assigned the task to keep these outposts replenished and to some extent protected.[6]

Initially, the Marines' DeHavilland biplanes were too small and their range too limited to be of much use. However, by August of 1928, the Marines put into service three tri-motor Fokkers capable of carrying as much as 1,300 pounds of cargo. Everything from cigarettes to mules were delivered by air and some of the more remote outposts received their payrolls in this manner.[7] To communicate with the aircraft in the absence of radio sets and receivers, pilots were instructed to interpret a prearranged display of cloth panels enabling a ground unit to "call for supplies, air support, medical assistance, or simply inform the pilot that there was no change in the situation."[8]

These methods proved extremely valuable in not only maintaining continual contact between remote units and headquarters, but perhaps for the first time in aviation history, the Marines pulled off an evacuation of critically injured men from the field and delivered them to medical facilities that ultimately saved their lives. This may have indeed been the precursor of the doctrine of air supply and support from the jungles of Central America to the jungles of Burma and ultimately to the jungles of Vietnam. In fact, in another instance in which amphibious planes were used to fly wounded out from a sandbar in the middle of a river, it was observed by both Marine and naval officers that a solution to the inaccessibility of the jungle environment called for a craft that had a high lift capacity and needed only a minimum of runway for takeoffs and landings. What they proposed was an aircraft that used rotary wings--a helicopter or an autogiro.[9]

Although the full practicability of helicopters would not be seen until much later, the development of such an aircraft continued. Due to the massive shrinkage of all United States forces in the two decades between the wars and the subsequent reduction in funding, it would not be until July 1940 that the Army Air Corps approved a contract to build the first experimental rotary wing aircraft, designated the Platt-LePage XR-1 twin rotor helicopter with the capability of carrying a 1,200 pound load. The delivery date of the XR-1 was to be March 14, 1942. The Air Corps had also contracted with Vought-Sikorsky for a single rotor helicopter, designated the Vought-Sikorsky XR-4, with a load capacity of 550 pounds. The delivery date for the XR-4 was January 10, 1941, at a cost of $50,000 per aircraft.[10]

By June 18, 1942, the United States Army Air Force (USAAF; its designation changed from Corps to Air Force in the spring of 1942), following flight demonstrations at Wright Field, had determined that the XR-4 was the helicopter most likely to be accepted in the future.[11] In fact, the USAAF was enthusiastic enough about the XR-4 to give the go-ahead to Vought-Sikorsky to build a slightly more powerful version of the XR-4, designated the XR-5, at a total cost of $650,000.[12] It was determined that the helicopter would be useful for many purposes: convoy duty; coastal and harbor patrol; observation and fire control; liaison and communication; ambulance duties; and, with the installation of low pressure floats in place of wheels, the helicopter would be able to land on water, marsh land, snow, or ice.[13] Apparently the authorities in Washington had enough insight to realize the possibilities of the helicopter, but perhaps what was needed was the appropriate environment to test the true potential of these rather peculiar looking aircraft.

During the 1930s, the Air Corps was also looking into development of various other types of aircraft, specifically for the ferrying of troops and/or supplies. In response to the success of the Germans in Europe in 1940 and 1941, the Air Corps began to consider the training of pilots to fly gliders. This training would provide a pool of pilots specifically trained "in handling gliders having flying characteristics similar to those required of troop carrying gliders, rather than the characteristics of the soarers [sic] used by glider enthusiasts."[14]

By June of 1941, the U. S. Navy had begun to draw up designs for both a landplane glider and a seaplane glider and by July was prepared to accept contract bids from various manufacturers. However, the Navy did not yet have plans to begin any sort of glider pilot training.[15]

In July, 1941, the Assistant Secretary of War for Air recommended that a conference be held for the specific "purpose of discussing the problems of loading and stowage in cargo type aircraft (that apparently did not include gliders) of certain articles used by parachute troops, air-borne Infantry, air-borne artillery, Engineers, and other units of an air-borne Task Force."[16]

It was not until November, however, that a proposal by the Assistant Secretary of War for Air called for a conference "of all interested Arms and Services, to determine future requirements for parachutes, for personnel and aerial delivery."[17] This conference would then determine "the basis for specifications for future development, comment on the suitability of existing designs, and discuss quantities for future procurement in order to avoid future shortages.[18]

The Waco Aircraft Company was contracted to construct "two 15-place gliders, which will be capable of carrying internally the standard jeep and a crew of six men."[19] In fact, the Air Corps had discovered that with only a slight alteration the 15-place glider could be capable of carrying: "a.15 troops or, b. A standard jeep and 6 men or, c. A 75MM. Howitzer."[20] The glider was apparently destined to play an integral part in the overall scheme of things for the Army Air Corps. One of the questions that remained was just exactly what that part was going to be.

Believing that in the near future an airborne assault of some sort was inevitable, the Chief of the Materials Division of the Air Corps submitted a listing of the available containers that were on hand for aerial delivery in January of 1941. There were three types of containers to deliver both dry and wet payloads in the field. These containers, Types A-3, A-4, A-5. equipped with either a standard silk or cotton parachute, were capable of delivering safely to the ground all sorts of liquids such as water and gasoline, as well as such fragile items as rifles and radio equipment.[21]

To enhance the Air Corps' flexibility, studies were conducted on the possibility of using either B-17 or B-24 bombers as tow planes for gliders. At first glance, using recently developed heavy bombers as tow planes would seem to be a waste of resources, since strategic bombing was the only available tool to the Americans to hit back at Germany months before our ground forces would see combat. However, what the planners had in mind was the possibility of using these heavy, four-engine bombers to tow three gliders simultaneously.[22] What was apparently not considered was the possibility of controlling three ungainly and difficult to fly aircraft without causing a possible three way mid-air crash, and also the loss of the bomber. In addition, this memorandum had requested information available from other countries on gliders and glider training.[23]

Throughout 1941 and into early 1942, the United States Army Air Corps was only just beginning to evaluate numerous ideas and theories on some sort of mobile air warfare to be used in conjunction with the disposition of ground troops from one point to another. In response to a letter to President Franklin Roosevelt about the head start that Germany enjoyed in the training of glider pilots,[24] the Assistant Secretary of War for Air responded on June 15, 1941 by saying that, "The Air Corps believes in the efficacy of glider-borne troops and is conducting a glider program to this end. Also, the Air Corps is in the process of training military personnel to operate cargo type gliders Consequently, it is believed that there exists in the Air Corps, adequate development, procurement, and training organizations to satisfy its likely glider needs."[25]

The Air Corps throughout 1941 had been conducting experiments and developing ideas that might prove useful at some later date if war broke out. As far back as the summer of 1940, the Armored Force Board at Fort Knox had been directed to commence studies followed by actual tests of the dropping of supplies by parachute. And by the summer of 1941, the net results of those tests and studies was the establishing of the practicability of supplying ammunition, fuel, food, and water by airplane using burlap parachutes and without special packaging."[26]

When war broke out, the armed forces of the United States no longer had the luxury of studying theories over the long term. To emphasize that point, in a letter of January 30, 1942 to Major General Walter R. Weaver, George H Davis, (an aircraft designer during WWI and after, then an investment banker from NYC) stated, "There is no time in this war emergency to do experimental work on improving the design of glider trailers,"[27] or even the planes that pulled them. Since the war was on, now was the time to decide exactly what equipment the Air Corps required, and how to maximize its use.

The equipment on hand by February 1, 1942, was essentially that which the Air Corps would be using throughout the war. Not only did the United States have the best cargo plane available, the Douglas DC-3 (designated by the military as the C-47, but it also had the heavier twin-engine C-46, the four-engine C-54 and the C-87. To do light transport work and to carry one or two passengers, the Air Force had liaison planes.

These light liaison planes had been undergoing flight tests in conjunction with Army maneuvers as early as April of 1941. Although these aircraft were slow, lacked any armor protection, and carried no weapons, their slow speeds and maneuverability enabled them to get in and out of tight spots that larger aircraft were unable to penetrate. Perhaps their vulnerability would prove to be a major liability in Europe (but it was not known at that time exactly what the weaknesses and strengths of the aircraft were in actual combat conditions), but in an environment such as the jungles of the Pacific islands or of Burma and Malaysia their seeming maneuverability might prove advantageous.

Based on the satisfactory operation of these aircraft during trials, the Assistant for Procurement Services recommended that the Army procure at least a dozen of them to be made available to the Air Force Combat Command. Following delivery of the aircraft in September 1941, a number of problems surfaced, including poor vision backwards and upwards, but it was determined that at minimum some two hundred of these planes be ordered as soon as possible. Piper Company, Aeronica Company, and Taylor Aircraft were chosen as the contractors for the Air Force's liaison aircraft. Shortly thereafter, Lieutenant General Delos C. Emmons (Deputy Chief, Army Air Corps) recommended that some 1,000 of the aircraft be produced. But because no funds were available (and indeed the Air Corps was running a deficit of over $300,000,000 by November of 1941), the enthusiastic endorsement of General Emmons was drastically altered by budgetary realities. As of November 10, an order was issued for eighty liaison aircraft with Taylorcraft given an order for twenty, Aeronica obtaining an order for twenty also, and Piper obtaining 50% of the order with a request for the delivery of forty aircraft.[28]

As the United States mobilized in late 1941 and early 1942, many aspects of the use of the resources available to the Air Force still required consideration. Major General George C. Kenney indicated in a letter to General Arnold (USAAF commander) dated April 25, 1942, that as troops moved forward or were placed behind enemy lines "they will receive fuel, ammunition, crews, food, etc., brought in by air."[29] Yet, what was left unanswered at this critical juncture was how this was to be accomplished most efficiently, with the least amount of risk. There existed the aircraft and the personnel to do the job, but it was still necessary to create a system of air supply and support.

Less than two months after Kenney's letter to Arnold, the War Department had released to the public, for the sake of public relations, the results of the launching of an airplane by a new "Air Pick-up System." A demonstration was held at Wright Field indicating the feasibility of a non-stop glider pick-up system. The glider was placed behind two uprights some two hundred feet back, attached by a towline. The tow plane approached the pick-up point in the same manner as if the pilot were about to land his plane. Reducing his airspeed to between 95 and 120 miles per hour, the pilot leveled off as he came near the uprights. With his pick-up arm lowered and the hook at the end fully extended, the pilot glided in toward the suspended tow line at a height of between twelve to fourteen feet. "At the moment of contact ... the cable reel inside the plane is permitted to spin freely to pay out additional tow-cable to cushion the initial load imposed by the dead weight of the glider on the ground. Gradually the reel-brake is applied, the glider accelerates smoothly, and by the time the speeding tow-plane has leveled off, the glider is airborne."[30]

This successful demonstration used a Stinson light monoplane as the pick-up plane and an XTG-3 as the glider. Although these planes were not meant for combat use, Saturday, May 30, 1942 marked the first time in aviation history that the theory of picking up light aircraft from relatively inaccessible areas by aircraft in flight became a reality. In fact, this operation may be considered the point of departure from theory to practice. Although the test took place in optimal, non-combat conditions, the demonstration proved the efficacy of placing troops practically anywhere conditions warranted, and then the ability to remove those troops when their mission had been completed.[31]

To accentuate that point, on July 25, 1942 Colonel W.D. Eckert (Executive for Control, Production Division, Air Corps) wrote a letter to the Chief of the Production Division of the USAAF about plans for the glider program. Based on projections for the coming year, the Air Force had been allowed to authorize the construction of 6,400 gliders, whereas projections for the numbers of gliders needed for the period from January 1, 1943 to June of 1944 were in excess of 41,000. And this projection was based on the premise that the gliders would be towed by either four-engine bombers or four-engine transports. However, if twin-engine medium bombers or transports were to be used, an additional 10,800 gliders was called for.[32]

Colonel Eckert had failed to realize that the heavy bombers of the United States Army Air Force, the B-17 and B-24, were not going to be allowed to tow gliders along with the rest of their duties. However, what the Air Force would have available would be the workhorse of cargo transport and the towing of gliders throughout World War Two: the C-47. In conjunction with the C-47, Colonel Eckert expanded his thoughts on the subject to include the usefulness of gliders. He indicated that gliders should be used as: 1) Full troop-carrying gliders for combat purposes; 2) As cargo-carrying ships between advance bases and permanent bases; 3) As personnel transports; and 4) As hospitals for evacuating the wounded."[33] Colonel Eckert summarized the versatility of gliders and the extent of their uses within the framework of a war. Not only could gliders be used in an offensive nature but rather than leaving them idle once their precious cargoes had been debarked, gliders could provide the means for perhaps the fastest and safest ways of getting seriously wounded personnel away from the battlefield to a properly equipped surgical hospital before the men would succumb to their wounds.

By April, 1942, the United States Army Air Force contracted with a number of manufacturers for the production of gliders. Companies as diverse as the Ford Motor Company, using their own designs based on parameters provided by the Air Force (1,000 gliders, model CG-4A at a cost of $18.9 million plus 15% for spare parts), Cessna (1,000 gliders designated the Waco model at a cost of $16,962,500.00 plus 15% for spare parts), the AGA Aviation Corporation (120 units at a cost of $2.2 million plus 15% for spare parts), the Babcock Aircraft Corp. (102 gliders, model CG-4A at a cost of $1,887,503.88 plus 15% for spare parts), and the Gibson Refrigerator Company (an order for 363 Waco CG-4A gliders at a cost of $6,912,128,75 plus 15% for spare parts) were given orders for gliders.[34] It was obvious that gliders were going to play a crucial part in the war effort. The only question was exactly how these carriers of men and materiel would be used to best advantage.

To complement the inventory, the development of the helicopter continued. These aircraft by the spring of 1942 were just emerging from the experimental stage and their practicability in a combat support role had not yet been proven. However, given the possibility of men caught in an inaccessible remote area with no way of returning to base, especially if they were wounded, a helicopter might be able to reach them when other aircraft and ground units could not. One thing was certain: the helicopter was virtually an unknown quantity before the shooting started. How it would perform under combat conditions and just how vulnerable it was to ground fire and enemy aircraft were still unanswerable questions.

In July of 1942, a conference was held including the commanders of the Army's G-4 (Supply Section), SOS (Services of Supply), ASC (Air Service Command) and RGS (Readiness Groups) sections (respectively, Colonels Walker, Berry, Newberry and Lt Colonel Generous) on the "Responsibility for Air Supply to Isolated Groups in Combat Zones."[35] It was crucial for the Army Air Force (AAF) to make that determination as soon as possible. The members of the conference reached a consensus that Troop Carrier Command was the unit best equipped and trained to deliver supplies to isolated groups (they believed that Air Transport Command was not properly equipped to drop supplies while in flight). In addition, the conference decided that the ground forces should be responsible for having their supplies made available for delivery by air to airborne troops and isolated ground units." Furthermore, the ground forces were responsible for ensuring that their supplies were delivered to the airdromes for the Troop Carrier planes to pick up.[36]

With that particular responsibility established, the Combined Joint Chiefs of Staff had to determine where and when these particular tools of war would be used. Making the vital decisions about allocation of scarce resources in very diverse theaters of war was not an easy task. The Allies had officially determined that Nazi Germany was the most dangerous threat; therefore, the defeat of Hitler was paramount. The Japanese, although at the threshold of India, and having extended their East-Asia Co-Prosperity Sphere by virtue of their apparently successful attack on Pearl Harbor[37] and subsequent occupations of Wake Island, Guam, and the

Philippines, would be dealt with fully by the Allies once the Germans were defeated. For the time being, the conflict against the Japanese would be waged primarily by the United States, followed by the Commonwealth troops of the British Empire and segments of the Chinese Nationalist Army. Whatever men and materiel could be spared from the Allied buildup against Germany would be sent to the Pacific and what few units that were left would be allocated to the China–Burma–India Theater of Operations (CBI).

To prepare an offensive in the CBI, the Allies required massive supply shipments from the United Kingdom and from the United States. These supplies were meant to keep Generalissimo Chiang Kai-Shek placated and keep his armies in the field opposing the Japanese. The Allies were compelled (they also intended to develop bases in eastern China to bomb Japan; therefore, the plans for air operations included maintaining these bases with the help of Chiang's forces) to provide thousands of tons of supplies per month to China that could be delivered only via an air route over the Himalayas. This commitment, in turn, used men and equipment that would otherwise have been better employed elsewhere. However, the Generalissimo was considered a vital component in the plans to defeat Japan. Therefore, short of placing troops on Chinese soil, this lesser of two evils was deemed the most desirable. Not until the Combined Joint Chiefs of Staff made their final plans for the future conduct of the war and obtained approval from President Roosevelt and Prime Minister Churchill would the Allied Command in India have any indication about their exact status in the overall war plans. Then and only then would they be able to determine what would be available and when, and, therefore, be capable of developing an offensive strategy against the Japanese.

Not only would the Allies need to adapt to a precarious 12,000 mile long supply line, but once the supplies reached India, the primitive nature of India's infrastructure was a further obstacle. Faced with inadequate port facilities, no electricity, running water, or all-weather roads, over-burdened rail lines that were also in disrepair, and deplorable conditions in the state of Assam in northeastern India, the Allies faced significant challenges prior to any offensive taking place.

Compounding their tribulations, by the middle of 1942 the command structure was in chaos. It was extremely difficult to determine who had what responsibility over the forces that were gathering. Commonwealth forces from all over India, divisions raised in both East and West Africa, Canadian units, Home units, United States forces (primarily USAAF), the American Volunteer Group in China (popularly known as the Flying Tigers), Chinese units under the command of Lt. General Joseph Stilwell, and those under various Chinese commanders, some Burmese units and some Australian and New Zealand units–all had to be brought under some sort of unified command.

As the basic command structure was being established, the entire components of cargo and carrier air groups had to be created in the United States and shipped to the theater of operations. The logistics of such a move were daunting in the extreme, considering the distances and the primitive nature of the facilities available in India. For example, the number of planes, primarily C-46s and C-47s, called for in a combat cargo group was 125. The number of crews needed for the group, including a reserve element, was 150 (a crew consisted of 4 men). In addition, the maintenance and ground support units and the necessary administrative and support personnel brought the total number of men to 883, of which 350 were officers.[38] This aggregate total of men and machines made the task of transportation an exceedingly difficult undertaking.

Lacking the necessary training for the job that the high command conceived for the air units, the Army indicated that it knew of no "diagrams or blueprints pertaining to loading and unloading methods of any transports in Army use."[39] It was however, common knowledge that the heaviest items were loaded as close to the wings as possible, whereas the lighter items were then distributed throughout the remaining space. If it was at all possible, items should be lashed down or secured by some sort of restraint. The rest were to be braced against the more firmly fastened supplies. All members of the crew participated in loading the cargo with the invaluable assistance of native labor. Not until the middle of 1944 was a loadmaster provided to expedite the loading of planes with little intercession on the part of the crew.[40]

The United States Army Air Force in 1943 was about to embark upon a mission unprecedented in its previous experience, and it certainly was going to encounter difficulties unlike those in any other theater of operations. Considering the lack of experience of the plane and ground crews, it remained to be seen if it would be possible for them to adapt their ingenuity and skills to fulfill the needs of the Allied campaign against the Japanese in Burma. Or would these men and their machines, heading into such a vast unknown, be crushed by all the varied physical, environmental, political, and logistical barriers set before them?

# Notes

1) Edward Meade Earl, ed., *Makers of Modern Strategy* (Princeton, NJ: Princeton University Press;1943), 499. It was also mentioned more specifically in the book, *Billy Mitchell: Crusader for Air Power* by Alfred F. Hundley, USAF, (Bloomington and London: Indiana University Press; 1975), 36., that Mitchell had submitted a plan to General Pershing in October, 1918 suggesting that the First Division, having the most experience in the Army be given parachute training for a drop behind the German lines.
2) Peter Calvocoressi and Guy Winty, *Total War* (Middlesex, England: Penguin Books; 1972),160. Over a third of the airborne troops were killed or wounded and the Luftwaffe lost some 220 aircraft. However, the Germans captured Crete and denied the British a base within range of the Ploesti oilfields. Plans for a similar attack on Malta were scrapped and Hitler turned all his parachute regiments into infantry units, destined for the Russian front.
3) Donald H. Connolly, Administrator of Civil Aeronautics to Brig. Gen. Oliver P. Echols, Acting Chief of Air Corps, 10/23/41; in File #454.1 "A," Box #753, Record Group 18, National Archives (Nat'l Archs), Washington, D. C.
4) Both of Cornelius Ryan's books, *A Bridge Too Far* and *The Longest Day* deal extensively with the airborne operations and their respective shortcomings. Also, Carlo D'Estes' *Bitter Victory* describes in great detail the problems that the Allies encountered in the invasion of Sicily. For the most recent account of the airborne operations in conjunction with the Allied invasion of Normandy, Stephen Ambrose's *D-Day June 6, 1944: The Climactic Battle of World War II* provides a great deal of information of the difficulties encountered by the airborne units not only during the initial drops and landings, but more importantly, the significant limitations that the US and British units had to function with and at the same time required to carry out their respective missions.
5) "Proposal to Build Helicopters" sent to the Office of the Chief of the Air Corps, June, 1924. File #052.18, Box #510, RG 18, Nat'l Archs, Wash., D.C.
6) *Marine Corps Historical Reference* Series, Number 21, "The United States Marines in Nicaragua," Historical Branch, G-3 Division, Headquarters, U. S. Marine Corps, Washington, D. C., Revised, 1961, pp 25-55.
7) Ibid.
8) Ibid. In one particular incident, Lieutenant Christian F. Shilt flew in a total of 1,400 pounds of medicines and supplies to a Marine unit in the small village of Quilali that was completely surrounded by Sandinistas. Singlehandedly, Shilt flew 18 men out, three whose wounds were of such serious nature that immediate medical attention meant life or death. In recognition of such a singular feat, the Marine Corps awarded Lt. Shilt the Congressional Medal of Honor.
9) Macaulay, Neil, *The Sandino Affair* (Chicago, Illinois: Quadrangle Books; 1967),124-125.
10) Headquarters, Army Air Forces, "Routing and Record Sheet," 6/18/42. File #452.11 "B," Box #753, p.1, RG 18, Nat'l Archs, Wash., D.C. This document relates to the status of Rotary Wing Projects.
11) Ibid, 2.

12) Ibid., 2.
13) Inter-Office Memorandum, War Department, Air Corps to Commanding General, AAF Materiel Command, 05/08/42, p.2. File #452.1 "B," Box #753, RG 18, Nat'l Archs, Wash., D.C.
14) Administrator of the Civil Aeronautics Adm. to General Oliver P. Echols, Acting Chief of Air Corps, 10/23/41. File #452.1 "A," Box 753, RG 18, Nat'l Archs, Wash.
15) "Report on Glider Training," Materiel Division to the Army Air Corps, 06/18/41. File #353.9, RG 18, Nat'l Archs, Wash.
16) Commander, Air Corps to Chief of Armored Force, 07/05/41. File #337.0, Box 1890, RG 18, Natl Archs, Wash.
17) F. I. Ordway, Jr., Major, Air Corps, Asst. Exec., Materiels Div. to the Chief of Infantry, 11/19/41. File #337.0, Box #1890, RG 18, Nat'l Archs, Wash.
18) Ibid.
19) Acting Chief, Air Corps to Air Staff, 12/10/41. File #452.1 "A," Box #753, RG 18, Nat'l Archs, Wash.
20) Ibid.
21) "Information on Air Corps Equipment," Materials Division to Intelligence Division, Army Air Corps, 01/23/41. File #452.3, Box #803, RG 18, Nat'l Archs, Wash.
22) Memo, the Office of the Chief of the Air Staff to the Chief of the Army Air Forces , 07/19/41. File #452.1 "A," Box #753, RG 18, Nat'l Archs, Wash.
23) F. I. Ordway, to Mr. Charles T. Malone, President, American Glider Association, 08/26/41. File 342.1 NA," RG18, Nat'l Archs, Wash.
24) Mr. Edward Steptoe Evans to Franklin Delano Roosevelt, 06/05/41. File #452.1 Box 753, RG 18, Nat'l Archs, Wash.
25) Robert A. Lovett. Assistant Sec'y of War for Air to Mr. Edward Evans in response to Edward Evans in response to his letter to FDR, 06/15/41. File #452.1 "A," Box 753, RG 18. Nat'l Archs, Wash.
26) Letter, "Armored Force Board Project No. 37, Supply by Air," Armored Force Board to the Army Air Corps, 01/28/42. File #452.3, Box #802, RG 18, Nat'l Archs, Wash,
27) George H. Davis to Major General Walter R. Weaver, Acting Chief, Army Air Corps, 01/30/42. File #454.1 "B," Box #752, RG 18, Nat'l Archs, Wash.
28) War Department, Office of the Chief of the Air Corps (from the Assistant for Procurement Services, General 0. P. Echols) to the Chief of the Army Air Forces, 02/04/42. File #452.1 "A," Box 754, RG 18, Nat'l Archs, Wash.
29) Letter, Major General George C. Kenney to General H. H. Arnold, Chief of the Army Air Forces, 04/25/42. File #385.C, Box #597, RG 18, Nat'l Archs, Wash.
30) Newsletter, "PICK-UP," All American Aviation, Inc., June of 1942. This document describes the first pick-up of a glider by an aircraft in flight. File #452.1 "B," Box #752, RG 18, Nat'l Archs, Wash.
31) Ibid.
32) Memorandum, "Planning of Glider Program, "from Colonel W.D. Eckert, Executive for Control, Production Division to the Chief, Production Division,

Army Air Corps, 07/25/42, p.4. File #452.1 "B," Box #752, RG 18, Nat'l Archs, Wash.

33) Ibid., 2.

34) Routing and Record Sheet to the Chief, Army Air Corps, a summation of contracts that had been negotiated with various manufacturers of gliders, 05/02/42. File #452.1 "B," Box #752, RG 18, Nat'l Archs, Wash.

35) Memorandum from Headquarters, Army Air Forces concerning the "Supply of Air Borne Troops & Isolated Units," to Colonel Whitten dated 07/20/42. Located in File #400.Misc., Box #215, RG 18, Nat'l Archs, Wash. 07/26/42.

36) File #400. Misc. Box #215. RG 18, Natl. Archs, Wash.

37) There is no question that at least on the face of it, the Japanese strike against the primary U. S. naval base in the Pacific was an enormous disaster both militarily and psychologically to the United States. However, upon closer examination, once all the fires had been doused and the dead and wounded taken care of, the situation at Pearl Harbor was far from the debacle that most people had come to believe. As Gordon Prange pointed out in his signal account of the attack, *At Dawn We Slept*, the final assessment at Pearl turned out to be greatly opposed to the general beliefs concerning the Japanese attack. There is of course no way to minimize the loss of some 2400 lives, including over 1100 men permanently entombed in the bowels of the *Arizona* and some 1100 wounded in the attack. Also, the sinking or seriously damaging of six of the eight battleships plus a dozen or so other ships sunk or damaged plus the destruction of some 188 aircraft at Hickam and Wheeler Fields certainly can not be ignored.

What the Japanese attack had accomplished seemed to have been a huge success in crippling the United States Navy's ability to interfere with their plans throughout South Asia and the western Pacific. But the Japanese, in fact, had missed a golden opportunity that would haunt them for the next 3 1/2 years. Yes, they did a great deal of damage and it seemed that they would now be unstoppable.

However, the Japanese had missed the three most important targets that would have indeed impeded the United States in any attempt to stop Japanese aggression. Their naval aviators left the fuel depots completely unscathed, therefore keeping the reach of the US Navy some 2,800 miles closer to the action instead of operating from the west coast. Also, the aviators missed the major repair facilities, which also meant that ships needing major overhauls or repairs would not necessarily have to go back to San Diego or Bremerton, Washington. This would be pointed up by the fact of the seriously damaged aircraft carrier, the *Yorktown* after returning from the Battle of the Coral Sea in May was repaired in record time for battle in the crucial Battle of Midway instead of being sent back to the west coast.

In addition, six of the eight battleships were all repaired at Pearl and eventually sent back to active duty. Plus the fact the battleships were in Pearl for the sole reason that they were too slow to keep up with our carriers meant that for all practical purposes, the battleships had become obsolete in the newly emerging warfare at sea using the speed and striking power of a navy's carriers and its aircraft.

And last, the Japanese had hoped to find two or three of our carriers sitting quietly at anchor on that peaceful Sunday morning. Needless to say they were sadly disappointed and to add salt to the wound their attack made the vast majority of the American population angry and promising to return the favor in spades

38) Wesley Frank Craven and James Lea Cate, *The Army Air Forces in World War* Two, Volume VI (Chicago, Illinois: University of Chicago Press; 1955), 391-2.

39) Memo, "Transport Loading," from Headquarters, Army Air Forces to Naval Air Transport Service, Office of the Chief of Naval Operations, 04/23/42. File #500—523, Box #887, RG 18, Nat l Archs, Wash.

40) Interview with Walter L. Carre. Mr. Carre originally served with the 7th Bomb Group as a mechanic and was transferred to a special rescue unit with their own pilots and aircraft comprising liaison aircraft, two B-25s and the first operational helicopter in the theater. Interview took place on 5/21/93 at King of Prussia, Pennsylvania.

# CHAPTER TWO

# "GO TELL THE RAF"

By late 1942, the outlines of Allied strategy against the Axis took on a more concrete and optimistic format. It was at the Casablanca Conference in January of 1943 that the "broad outlines of the strategies" of the Allies took shape. The Mediterranean was made the principal theater of operations in Europe, to be followed by an invasion of northwest Europe late in 1943 or early in 1944. On the other side of the world, operations in the Pacific theater would lay claim to most of the resources (primarily American), for securing the Aleutians, advancing from Midway on the Truk-Guam line with a concurrent advance from Rabaul if forces were available, and advancing via Samoa and the Malay barrier. Support for activities to secure Chinese participation in Burma under the code name of "Operation Ravenous" would be limited to a small advance from Assam to obtain jumping-off points for further advances into Burma at some future date, and to improve the air-transport route (the "Hump") to China. Any ground campaign would be sufficiently supplied, but there would be little available in the way of assault and landing craft, naval forces, and shipping.1 In other words, the China-Burma-India Theater of Operations (CBI) would be given some of the tools needed to conduct some sort of offensive against the Japanese.

Adding to these difficulties of priorities, the topography of Burma, coupled with its unique weather conditions, created hazards and problems on a significantly larger scale than encountered by the Allies elsewhere in the world. Burma, which was to become a battleground for Chinese, Americans, British, Indians, Ghurkas, East and West Africans, and of course Japanese for more than three and one-half years, could not have been a more horrendous theater to fight a war.

Almost the size of Texas, Burma has been likened to a giant wedge driven between China and India. A massive ring of mountain ranges which form most of its 3,200 miles of land frontier locked the country in on three sides. And prior to completion of the Burma Road. there was no rail or road connection with China or India. The Patkai-Naga Hills, with peaks rising to approximately 20,000 feet and valleys of impenetrable jungle, interposed an effective bar to transportation between Assam and northern Burma.

Farther to the southwest, an extension of that range, the Arakan Yomas, stretched to the shores of the Bay of Bengal, with peaks up to 10,000 feet, preventing land communications between India and lower Burma. On the east, too, Burma had been long isolated from its neighbors by rugged and forbidding mountains. On the China-Thailand side, the Himalayan range branched out into numerous parallel north-south chains, extremely high along the China border, tapering gradually in the Shan states, and stretching far down into the Malay peninsula. In shape and contour, Burma itself resembles a slightly cupped hand, a jumbled mass of parallel ridges running north and south giving the palm a corduroy-like appearance.

The central basin, its ridges shrinking in altitude toward the middle, extends from Fort Hertz in the north to the Bay of Bengal, varying in width from 100 to 150 miles. Because the north-south ranges hinder lateral traffic, the valleys between have developed into heavily traveled routes. Upon the southward-flowing rivers, most important of which is the Irrawaddy-Chindwin, and upon motor roads and rail lines closely paralleling their courses, normally passes the bulk of freight. After the victory of the Japanese in 1942, this system became the lifeline of enemy occupation forces.[2]

Burma's tropical and sub-tropical climate further compelled the opposing forces to alter their plans of campaign. With rainfall exceeding more than 120 inches in some parts of the south and monsoons blowing in regularly between May and October, the conduct of ground and air operations was severely hampered. Only during the "dry" season of October through May were the Allies able at first to commence offensive operations, and that is not to say that it did not rain from November through April. Indeed, as Lt. General Joseph Stilwell noted in his diary during the 1943-44 campaign, "It rained about eleven inches in January and another eight inches more in February."[3]

If the rainfall was not enough of a hindrance, the maximum temperatures at Shwebo and Mandalay in central Burma reached above 100 degrees Fahrenheit, whereas farther north in Myitkyina the temperature never went below 98 degrees.[4] In such a climate the prevalence of parasites such as leeches and mosquitoes causing such debilitating diseases as malaria, bacillary dysentery, and cholera hindered the fighting capabilities of all the forces in Burma.

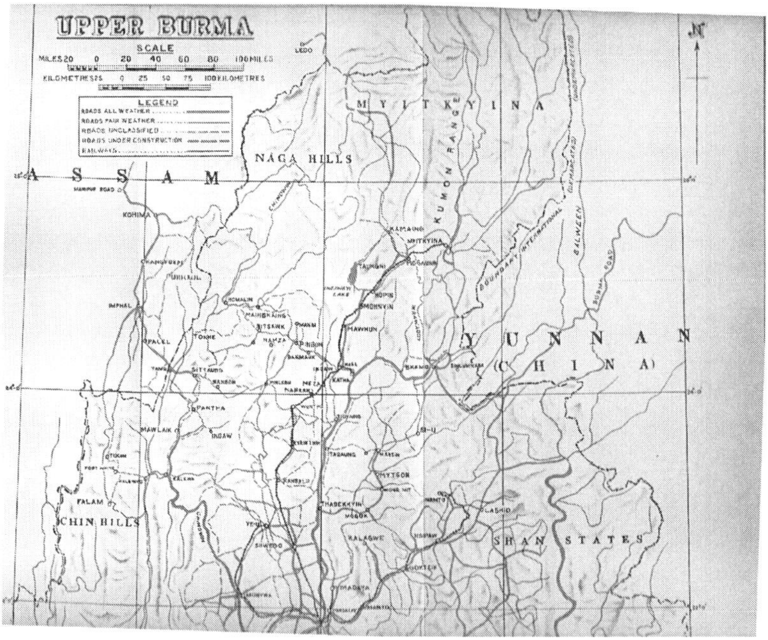


Picture courtesy of Indian Armed Forces in WWII – Reconquest of Burma Vol 1.

Given all these obstacles in the forward areas along the Indian-Burmese frontier, the primitive nature of the infrastructure in the rear areas needed consideration long before any Allied offensive was to occur. In the Assam region alongside the state of Bengal (which bordered on the northwest region of Burma), any Allied buildup would require essentially starting from scratch. Airfields, hangars, barracks, fuel depots, training facilities, ammunition dumps, base hospitals, and supply dumps for everything from automobiles to Zippo lighters had to be constructed. Very few of the supplies could be left exposed to the heat, humidity, and rain that plagued the region. Therefore, the tasks assigned to the various Allied units upon their arrival in the Assam region initially fell into the category of housekeeping chores long before their air and land assaults against the Japanese could begin.

By December of 1943, the Eastern Air Command had been established with its commander, Major General George E. Stratemeyer, who would be answerable to the commander of South East Asia Command, Admiral Lord Louis Mountbatten. The Eastern Air Command consisted of the 10th Air Force Strategic Air Services, Troop Carrier Command, and a number of Royal Air Force Groups. Within the Troop Carrier Command, Brigadier General William D. Old was ordered to "supervise all Air Transport operations in the Forward Areas."[5] Troop Carrier Command had the 1st Troop Carrier Squadron (TCS); the 2nd TCS; the 27th TCS; 315th TCS; 31 Squadron, Royal Air Force (RAF); 62 Squadron (RAF); 117 Squadron (RAF); and the 194 Squadron (RAF).[6]

Airdrome (airbase) Squadrons were organized for "tactical employment as ground servicing units to support Combat Cargo"[7] and Troop Carrier Squadrons. These Airdrome Squadrons were assigned the jobs of "administration, mess and security, chemical and medical services, furnish and provide first and second echelon maintenance and supply and to organize and operate airdromes."[8] Essentially, the Airdrome Squadrons were to provide all the necessary ground support services required by both Troop Carrier Command and Combat Cargo Command; in the case of Troop Carrier Command the 98th Airborne Squadron was to serve as the housekeeping unit for the 27th and 315th Squadrons.[9] They had to assemble an efficient base of operations to handle the various contingencies of combat air crews and their aircraft.

General Old, having been named commanding officer of Troop Carrier Command on December 16, 1943,[10] was charged with the task of "providing air transportation for airborne and air transit forces in the support and training of the Army Group and other land or air forces included in operations in Burma."[11] This order would place Old's Troop Carrier Command in the forefront of providing, by any means available, all needed support of ground troops on the front lines and especially those placed some 150 miles behind the Japanese lines, as envisioned by British Major General Orde Wingate.

It was all well and good to have at his disposal on paper such a large air armada. Each U.S. squadron was provided with thirteen C-47 aircraft, "with provision for one and a half crews per aircraft."[12] Each Royal Air Force Squadron was equipped with twenty-five Dakotas, which had been modified with the addition of a number of .30 caliber machine guns for defense and a number of navigational aids.[13] Of its twenty-five aircraft, twenty were operational, with the remaining five held in reserve.[14] All told, General Old commanded in excess of some 150 first-rate transport planes to effect a mission never before required in the history of aviation. Could SEAC, through an Eastern Air Command that filtered all the way down to specific squadrons, succeed at a job that had never been attempted before? Old's command had to contend with a precarious supply line across the Pacific to the west coast of the United States. And to obtain supplies from Great Britain, the traffic had to endure some 6,000 miles of sea and land routes stretching from Assam across India and through the Middle East, across the Mediterranean and to England past the major submarine bases of the German Navy.

As a testament to the fragility of this supply line, items that were taken for granted in Europe or the Pacific were at times difficult, if not impossible to obtain in the CBI. If anything of value to frontline units happened to have arrived in the rear depots, a unit had to lay claim as quickly as possible to these items, for otherwise they would disappear in less than twenty-four hours. As Colonel Philip Cochran (commanding officer of Air Commando 1) related, "You'd send all this beautiful new stuff, and they would grab it. They would take everything you got as it would come into ports. They would steal it from you. An unalerted[sic] task force could lose their identity before they even got into place, because their supplies would be used by people who said, 'My God, we haven't seen any of those for 2 years. I think we need those.'"[15]

One of the most annoying problems for the South East Asia Command (SEAC) was the difficulty in obtaining parachutes. In this newly evolving concept of air mobility amid impenetrable jungle, the quickest and safest method of delivery was dropping all the requisite supplies out of C-47s, either by parachute, or in many instances, without. As noted by Mountbatten's staff, the need for parachutes in operations conducted by Wingate's "Chindits" (a unit named by the corruption of the Burmese word for the winged stoned lions that are the guardians of Buddhist temples) called for a quantity of parachutes that in late December of 1943 simply did not exist. The heat and humidity affected the silk material by spreading mold and mildew on the fabric. Therefore, the usual silk material needed another component to strengthen the fabric. With these facts in mind, it was recommended by SEAC that Operation Tarzan (Wingate's Long Range Penetration Group) wait until the parachute supply problem could be eliminated.[16]

To alleviate this shortage, SEAC devised a number of methods to increase its supplies on hand rather than wait for months for shipments from Great Britain and the United States. Some of those recommendations were as follows: local cotton supplies were to be used for parachutes more than ever before; a local material, jute, was tested as a replacement for silk or cotton; and it was determined that a larger segment of supplies needed by men and animals in the jungle could in fact be free dropped with little risk of damage. Items previously dropped with parachutes were reconsidered as being able to sustain a drop of some three hundred feet with little or no damage. Along with such items as forage for animals and food grains for personnel and clothing, other non-fragile items such as tinned food, which at present required either a parachute[17] or a similar speed-reducing device, could also be free dropped.

Additional experiments were carried out to increase the loads for which various parachutes had been previously specified. The British had in fact during the months of May and June of 1943 carried out experiments with the intent of increasing the loads that various parachutes could accommodate. Instead of cutting the parachute material on the bias, they determined that cutting the material on the square would enable the parachutes to not only carry a heavier load, but also increase the descent of the items only slightly with no undo harm to either the items dropped and on the parachute canopies themselves. In fact, of ninety-three drops made with parachutes cut on the square, only one parachute suffered a slight tear in one of the "A" panels near the periphery.[18] Therefore, it was determined that this method would enable the cargo planes to deliver more supplies to the forces on the ground not only quicker, but also with little additional risk of damage.

The Mark I parachute load of some 140 pounds had been increased to more than 180 pounds. The Mark II parachute had its carrying capacity increased from 180 pounds to nearly 210 pounds, and it was assumed that further improvements would raise that load to some 250 pounds. However, the use of jute for parachute material did not fare as well. Although designs were made that enabled an eighteen-foot jute parachute to carry a load of 200 pounds, its rate of descent (some 35 ft/second) was considered too high and forced the "parajutes" to be used for only non-fragile items that could not be free dropped.[19]

The experiments with the jute parachute proved to be disappointing. The weight of the "parajute" was thirty-three pounds, versus a weight of nine pounds for the U.S. cotton Mark I and eleven pounds for the Mark II. This would increase the weight so substantially that for every seventeen fully loaded C-47s, one C-47 would be needed to carry the additional weight of the "parajutes" alone. If that were not serious enough, humidity acted upon the jute material, causing the "parajute" to absorb moisture in the air, and subsequently increasing its weight by more than twelve pounds. Last, the "parajutes," unless hermetically sealed in storage, had only a two month shelf life before they began to deteriorate and become useless.[20]

However, making the distinction between supplies that were virtually invulnerable to harm upon impact and supplies that required the greatest of care, SEAC dove enthusiastically into the design and production of baskets with native materials. These baskets would not only be more economical, but they also would be used in place of the jute parachutes because it was not feasible to drop anything fragile with the jute parachute. Specifically, fodder for the mules and horses, tinned food containers for the men and also, surprisingly, petrol and oil were some of the items that used the baskets.[21] All that was needed from either Great Britain or the United States were the metal filler caps that were unavailable in the theater.[22]

As noted in Mountbatten's diary, the supply of parachutes on hand by early February 1944 was not sufficient to carry out the various operations planned before the monsoons. By February 1, including all twenty-eight foot, twenty-two foot and eighteen-foot parachutes and those that were due to arrive from overseas, SEAC had the sum total of 174,084 parachutes for operations in the spring. However, based on the plans for offensive operations in February and March, some 155,000 parachutes would be used but never recovered, leaving a grand total of fewer than 19,000 parachutes available for any additional operations that spring.[23] Bearing these numbers in mind, SEAC realized that it was essential to develop [the] manufacture of gunny parachutes"[24] as soon as possible, or the plans for Wingate's LRPG would be put in jeopardy.

In fact, SEAC had become nearly obsessed to procure as many parachutes from across the globe as feasible. Based on projections of cargo being dropped in the amount of some 3,500 long tons per month from January through the following April, the British Air Commission in Washington requisitioned 200,000 aerial delivery parachutes from United States production to be delivered to the CBI, commencing in December with delivery of some 40,000. According to SEAC, production of parachutes in Great Britain and India in the coming campaigns was woefully insufficient for SEAC's needs due to a lack of materials and productive capacity.[25]

As bases were being established in Assam in the late fall and early winter of 1943, SEAC issued a directive to all air force commands settling once and for all the chain of command of the combat units of the Royal Air Force and the United States Army Air Force. All units were now to be united in a single Allied Air Force under the command of Air Chief Marshal Sir Richard Peirse. All combat units of the 10th Air Force and the RAF Bengal Command would be integrated into one air striking force known as the Eastern Air Command, Major General George E. Stratemeyer commanding, and also designated as second in command of all Allied air forces in the theater.[26] This move would ultimately stop the bickering between the USAAF and the RAF and facilitate cooperation in the upcoming offensives.

To highlight the need for this reorganization, Lord Mountbatten, SEAC commander, explained, "The South East Asia Command has been set up as an Allied Command because experience in this war has shown that this is the most efficient setup for defeating our common enemy. An Allied Command works best with closely integrated forces and unification of our Air Forces is a logical step. I feel certain that the American and British Air Forces will prove to be far more than twice as effective and deadly now that they are united into a single allied Air Force than either of them operating on their own. The closer our forces draw together the quicker will victory be achieved."[27]

To emphasize the point even more, Major General Stratemeyer issued a General Order No. 1 to all RAF and USAAF commands less than a week after Mountbatten's proclamation: "A resourceful, able and wily enemy must be blasted from the jungles of Burma and driven from the skies in days to come. His lines of communication must be obliterated, his shipping destroyed, his will to resist crushed. Against the inevitable day of retribution when Japan's cities will meet the fate of Berlin our life line to China must be strengthened and protected. Every ounce of energy of every man of this command will be required to accomplish this purpose. We must merge into one unified force, in thought and in deed - a force neither British nor American, with the faults of neither and the virtues of both. There is no time for mistrust and suspicion."[28]

Stirring words indeed, but mere words alone would not erase the mistrust and suspicions typical of both air forces. As in any alliance, problems would exist merging two different doctrines and many theories on the proper use of combat aircraft. Could the Americans and the British manage to set aside all of that in the name of Allied unity? The proof would come when the air forces supported units in the field, and not on the basis of stirring words from either Mountbatten or Stratemeyer.

Not only would two very different air forces be compelled to work together while trying to fight the Japanese, but they also had to work out the flaws in the system of supply and support of ground troops by air. In addition, if they were to carry out Major General Orde Wingate's plans for another long range penetration mission deep behind Japanese lines, all air force personnel would have to acquire the skills necessary to keep Wingate's "Chindits"[29] properly supplied and supported by tactical air.[30] Once the multitude of USAAF units began to arrive in India by late 1943 and early 1944, it became all too apparent that regardless of the training they had received in the States, they had to be retrained for this new type of warfare.

Two diverse units exemplify the problems encountered upon arrival and the steps that needed to be taken to make them ready to support units in the field. The creation of Air Commando One by fiat of the commanding general of the USAAF, Lt. General H. H. Arnold, in September of 1943, left little time for the outfit to train as a unit stateside. Therefore, not until Lt. Colonel Philip Cochran and his entire unit arrived in India was Air Commando One able to begin training in the new art of close-in air supply and support.[31]

Herein lay the basics of an operation that had never been tried before. Instead of Wingate's Chindits walking through the jungle with all their equipment, provisions, and supplies, perhaps taking as much as a month to reach their designated position, they now would be flown in, prepare their ground positions as a stronghold with an airstrip to accommodate C-47s, and then move out from these strongholds to wreak havoc on the Japanese interior lines. If air superiority was maintained, then Air Commando 1 could fly in at will during daylight and maintain these ground units for as long as they were in the field.

But before any of this was to take place, the training of the air and ground personnel became the most crucial factor determining whether the theories could be translated into reality. Whereas Air Commando One entered India with little or no training as a cohesive unit in the very difficult art of close-in air supply and support, one of Air Commando's support units, the 27th Troop Carrier Squadron, came to the CBI with a great deal of training in this new art of warfare. In fact, when Colonel Cochran was scouring US bases for volunteers for his new outfit, he had specifically commandeered the 27th to be part of his command.[32]

The 27$^{th}$, created in February of 1942, had undergone extensive training with both cargo planes and gliders in the United States until it was ordered out to join Cochran's Air Commando One in November of 1943. In fact, it had achieved such an outstanding record as a training squadron that earlier in 1943 orders had been given to ship out for overseas duty, but they were twice overridden by superiors in Washington.[33] Following the unit's transfer to Dunnellon Field in Florida in the early Fall of 1943, the 27th had managed to achieve a superb record in training other units for air supply and support and its reputation attracted Cochran's attention.

While stationed at Dunnellon, the 27th not only trained men who would become the core of soon to be activated units, but they also demonstrated the feasibility of towing fully loaded gliders, the transporting of engineer units from one base of operations to another, the carrying of supplies and equipment, the evacuation of men needing immediate medical care, and the dropping of paratroopers and parapacks under realistic battle conditions.[34] They perfected many of the aspects of air supply and support and also provided a highly trained core of men who would provide the much needed experience called for in the future. Having trained in the United States prior to being shipped out, they had not only received practice in towing gliders, but also in dropping supplies to troops in the field, deployed in the Everglades to mimic conditions of a jungle environment.[35]

Having received orders for the CBI on November 30th, the 27th departed Florida on December 13th and arrived in Karachi on January 4, 1944. Unlike most other United States forces that arrived in the CBI, the 27th was an intact unit that commenced flying combat missions within twenty-four hours of arrival. To demonstrate its readiness for action, the entire squadron, upon its approach to the base at Sylhet, India, came in at treetop level in perfect formation. As a reward for such expert flying, General Old, who had been directly below them as the squadron approached, ordered the unit to prepare to fly missions at once.[36] And they were ready.

On the 13th of January, the 27th flew its first mission out of Sylhet by flying twelve C-47s to a British base at Argatela, India, where they were loaded with British supplies to be delivered to Allied forces that were fighting for their very survival at Tiddim, just north of the Arakan. Under the escort of twelve British Hurricane fighters, the 27th commenced parachute droppings that called for the C-47s to fly an extremely "tight pattern around a viciously situated objective, where the extremely rough air necessitated accurate judgement[sic] and a strong pair of arms and legs."[37] Regrettably, the 27$^{th}$'s efforts proved futile, and Tiddim fell to the Japanese, but it was regained a number of months later.

If nothing else, at this particular juncture, the 27th Troop Carrier Squadron (TCS) not only demonstrated the feasibility of air supply and support, but it also proved that given sufficient support in men and supplies, it was quite conceivable that even units encircled by the enemy could survive long enough for either rescue or when their mission had been accomplished.

One of the most hazardous aspects of the training concerned the delivery of fully loaded gliders and the subsequent pick up of gliders from makeshift airfields in the bush. Consequently, of the four planned Long Range Penetration Groups (LRPGs) in early 1944, it became necessary to trim back to three, since there were not yet available the aircraft and experienced crews to perform with little risk to the troops.[38]

Between early January and early March, an active period of training ensued to assure success for the upcoming operation, code named

"THURSDAY," which would take General Wingate's forces deep behind enemy lines. In particular, a segment of Air Commando One was designated to train in the dropping of supplies at night in coordination with ground personnel. Included in this training was not only the dropping of supplies, but also day and night takeoffs with CG-4 gliders (commonly known as Wacos), with single and double tow, day pick-ups (snatches of gliders) loaded and light, and night pick-ups of gliders."[39] In addition, gliders containing a wide range of equipment and supplies, including pack animals were used, enabling the pilots and crews to familiarize themselves with never before tried techniques.

As preparations were put into effect, General Old was surprised to find that his Troop Carrier Command would have to perform as many as thirty-six daily sorties in support of Wingate's operation.[40] However, the following day it was agreed that in the future all information on operations that required air support be made known to General Old and his staff in advance.[41]

To emphasize that point further, Old, upon returning to his headquarters on January 14, following three days of reconnaissance with a Wingate staff member, was notified that new demands would be placed on his TCC, including an operation by a parachute brigade dropped behind enemy lines requiring extensive use of transport aircraft and the dropping of supplies. This addition called for the relief of 62 Squadron from its supply-dropping assignment in support of British units participating in the Kaladan operation to join up with the 177th Wing, which had been assigned the task of complete air supply and support of the parachute brigade.[42] As it turned out, although this new development underscored the problems of direct communication between British and American units, General Old found it satisfactory to have the entire 177th Wing involved in this upcoming operation.

However, Old hastened to add that the Tiddim operation would be in jeopardy due to the withdrawal of his units, while adding that if he was not kept fully informed on any or all future missions that required his units, he would be unable to plan any future Troop Carrier operations.[43]

On January 16, 1944, General Old sat down with Colonel Cochran for the first time since their respective arrivals in India to discuss the "air phase of the Wingate Expedition."[44] They mutually concluded that if any of Air Commando 1's special equipment was needed by TCC for any rescue work, the equipment would be furnished by Cochran's unit.[45] This generous gesture by Cochran may well have been intended to smooth any ruffled feathers or egos because of the unique status of Air Commando 1. (See explanation later in chapter 4.)

The next day a staff meeting discussed the latest information about the supply dropping requirements for Wingate's expedition. As General Old put it, it was like the plan to fly a round trip from Richmond, Virginia, to New York City without the benefit of reliable support facilities, along with unpredictable weather: "Fuel requirements will cut down on loads appreciably, and the engine cost and demands on crew would be terrific. Maintenance and replacement problems will be greatly increased and the navigational expertness required to find dropping points would be a test of a veteran pilot. This is terrible planning."[46] His staff had incorrectly asserted that the job could be done based on the experience acquired during the first Wingate expedition in 1943. As it turned out, only one or two long flights had been made to resupply Wingate's Chindits, sorties that were from 150 to 200 miles shorter in air miles than the proposed landing sites for "Operation THURSDAY." In addition, they had not been successful. Therefore, the experience acquired in Wingate's first operation had little or no relevance to the plans being developed for "THURSDAY," since the operation did not in any way resemble what Old's units would be called upon to do.[47]

Just prior to Old's and Cochran's January 7th meeting, a group of transports and gliders were taken to Lalitpur, India, to commence a demonstration of glider pick-up and troop-carrying to both Wingate and Mountbatten. Under the command of Captain William Taylor, the pilots and crews from Air Commando 1 were to show the dignitaries what they had learned a week earlier at a base near the village of Gwalion, where new tactics in supply and food dropping and the dropping of agents behind enemy lines had been practiced. Wingate, being the most enthusiastic proponent of this new type of warfare, elected to ride one of the first gliders that was snatched by Taylor's C-47s. Just before his glider was lifted off the ground, he yelled out of the window to one of his aides, "Go tell the RAF that I have not only seen it done but that I am now doing it too."[48]

Obviously, Wingate was extremely pleased with the demonstration and it would not be extreme to conclude that his plans for the future solidified that day. To underline that point, following a maneuver in which a British column of more than three hundred men and their mules were loaded into gliders and then snatched off the ground by Air Commando 1 C-47s, Wingate exclaimed that, "This is what I've prayed for," followed by Mountbatten's comment, "Jesus Christ."[49] Their remarks indicated that the naysayers in the RAF who had advised Lord Mountbatten that such a technique was totally impractical had been overruled by the reality of Air Commando 1's successes.

Shortly thereafter, a call from the British was made to Air Commando 1 to assist in the evacuation of wounded men from the Arakan front. A contingent of liaison aircraft, including L-ls and L-5s (light planes with no armor or armaments, slow flying, and limited range) was flown down by their sergeants and was immediately sent to areas under direct enemy fire. Little strips hacked out of the jungle were the only means by which critically wounded soldiers could survive their wounds if immediate evacuation was necessary. With Japanese mortar rounds whizzing over their heads, the pilots brought their light planes to a short stop with the engines still revving and awaited the emergence from the jungle of a number of seriously wounded men. As recounted in the official history of Air Commando 1, "A man with a smile on his face and his guts blown out was tucked in the rear seat of the plane or in the stretcher if necessary, and flown a short distance at tree top level to the base where they [sic] could be taken in transports to rear hospitals. British Tommies, West Africans, and Ghurka troops were evacuated from the battle area and as they were delivered to the Doctors and Nurses, they humbly thanked the American lads who had saved their lives."[50] The use of these flimsy, unarmed aircraft would make the difference between living or dying for countless soldiers on the front lines in the Burma campaign.

As the plans for the air invasion of Burma were being established, there was still a great deal for the various USAAF units to learn and prepare for "Operation THURSDAY." The luxury of time was not available to SEAC since the monsoon season would most likely commence in early May, hampering all operations. SEAC, as early as the previous 27th of November, had indicated that by removing many of the planes and crews from their duties of flying the "Hump" from India to Chungking [Chongging], China, the units would need only two weeks of final training to prepare them for future operations in Burma. As they pointed out, "air dropping and the transport of cargo and personnel do not require the same degree of training as is required for those units which will be used in dropping the Parachute Brigade and supporting LRPG. Further, it is assumed that the latter functions will be performed by the RAF squadrons now in training for such purpose."[51]

Experts at least acknowledged that what the Air Force units were being detailed to do in the planned spring offensive was indeed something quite unique in the experience of any of the numerous U S. squadrons that were slotted for the massive Allied assault. In conjunction with the air assault of Wingate's "Chindits," SEAC was embarking upon a multi-pronged ground assault across a 700 mile front to dislodge the Japanese from the banks of the Chindwin River in the south, as well as assaults to bring

General Stilwell's Chinese units and his one and only American ground unit toward the key railroad hub and airfield, Myitkyina, in northern Burma. Also, General Slim's XIV Army was to attack across central Burma, with the largest city outside of Rangoon, Mandalay, as their objective. This massive invasion would call for a better method of coordination and cooperation between not only diverse Allied units, but also the units on the ground and their guardian angels in the air. Given the enormous size of the operation and the unprecedented tactics never before attempted, was it possible to achieve the entire operation without errors that would destroy SEAC's chances for success?

The key to success or failure would be laid directly on Wingate's shoulders and those of the commanding officer of the American ground unit, the 5307th Composite Group (commonly known as Merrill's Marauders). General Stilwell appointed one of his sons-in-law, Frank Merrill, to lead a unit of men (code named Galahad) who had volunteered to participate in a relatively limited and short-lived operation (according to the War Department, this unit was to have one mission of approximately 3 months duration, and they were expected to suffer approximately 859 casualties), and whose purpose was to provide experienced jungle fighters so as to block Japanese reinforcements from coming to aid their besieged comrades at Myitkyina. Because they were the only U. S. ground unit in all of Asia, it was imperative that they not be wasted or misused. Galahad, having been modeled after Wingate's Long Range Penetration Group, was created for the singular purpose of giving Stilwell an American unit of high reliability to enhance the chances of his Chinese units in their assault on Myitkyina.[52]

By the beginning of January, all efforts leading up to the spring offensive had been put into motion. The necessary Air Force units had by then arrived in India and commenced training operations and actual combat missions to hone their skills for the upcoming push. The 27th Troop Carrier Squadron (TCS) had arrived from the United States and the 315th Troop Carrier Squadron had been activated in Dinjan and moved to its forward base at Sylhet.[53]

Training intensified. Even though the 27$^{th}$ TCS had acquired a great deal of experience back in the United States, they also had to adapt their tactics to a new kind of warfare. As noted by General Old on January 26, the efforts by many of his squadrons had fallen short of expectations. In one incident, a ground observer attached to the West African brigade had noted numerous deficiencies in a recent air drop. Included in these criticisms were "the failure of aircraft to await ground signal, parachutes dropped from too great height, free drops made too low, too much unloading at same time, improper marking of containers. Result is that many packages landed in water, others badly damaged."[54]

Most likely, General Old was not surprised by the mistakes. It is safe to say that all previous experience was useful, but if SEAC was going to succeed in this operation, a great deal more coordination and hands-on experience was called for. In addition, the dropping of supplies and the pick-up of gliders at night were being introduced as part of TCC's training since the Japanese would most likely be unable to hinder operations then.

By late January, night training for the 27th and 315th Squadrons had begun, with satisfactory results.[55] It had also been determined that the loading of wounded at collecting stations had been inconsistent. In some instances it took less than twenty minutes to get the men on board and securely placed inside the aircraft, whereas in other situations it took as long as one hour and ten minutes to accomplish the task. (The wide variance in time depended upon how long the troopers took to load themselves and their gear securely in the gliders. The weather played an important part in either expediting the loading or slowing it down.[56])

Adding to the difficulties, a number of aircraft had been damaged while engaged in dropping operations, and numerous aircraft had been lost due to the weather, mechanical difficulties, and on occasion, enemy fire. In one encounter with enemy ground fire, the plane had managed to gain sufficient altitude to enable the crew to parachute from the aircraft with sufficient height for the parachutes to function properly. However, not enough detachable parachutes had been available to the crew, resulting in the survival of only two of the seven crewmen.[57]

Whatever other operations for TCC had been planned, Old's unit was expected to evacuate 270 casualties a day between February 15th and March 15th, followed by an increase to 300 casualties a day between March 15 and April 15, and then by an increase to 500 a day between April 15 and May 31. In response to this request, General Old determined that a maximum of 210 casualties could be evacuated each day provided that evacuations be coordinated with supply dropping operations.[58] Such expectations on the part of Old's superiors had to come in line with the reality of the resources at Old's disposal. Obviously, it was necessary for compromise, otherwise TCC would have been pushed beyond its operational limits, thereby endangering all operations.

As training and experimentation continued throughout the month of February, reports were being received by TCC about parachute altitude experiments conducted by Slim's XIV Army. Experts had determined that the use of the larger 28-foot parachutes capable of carrying a load of between 280 and 300 pounds required an altitude of at least 500 feet to function properly, whereas the normal accepted height for smaller parachutes including the 18-foot and 22-foot types, was known to be only 200 feet. Also, the higher altitudes substantially reduced accuracy and put the aircraft at higher risk of damage from ground fire (the jungle canopy provided cover to aircraft that flew at treetop level). To make matters worse, "a planeload of some 25 containers, each averaging between 280 and 300 pounds was found difficult to manhandle and it is doubtful whether more than two containers can be ejected on one run over the target. This would envolve [sic] 13 runs on a target to unload one aircraft and the increased hazard of such a prolonged run is obvious."[59]

All these difficulties and countless others had to be addressed and remedied. General Old, on February 7, flew over the selected areas for the operations scheduled for March to observe moonlight visibility. Along the way he had observed some Japanese motor transport activity south of Kamaing and dropped a number of 100 pound bombs without being able to assess the damage. Having noted that the moonlight provided excellent visibility, he headed back to base, only to be fired upon by "friendly" anti-aircraft fire from the town of Imphal.[60] Apparently, because Old's B-25 was undamaged, a protest was never made to the British.

At the end of February, an assessment of supply-dropping experiments was prepared by headquarters, TCC. By using a C-46 rather than a C-47, TCC was hoping to learn if indeed the larger but harder-to-fly aircraft was suitable for dropping supplies and also for use as a tow plane for Waco gliders. Four tests over a three day period were conducted: the first test on February 10th was used to determine the flying capabilities of the C-46 in a simulated combat condition without any cargo in its hold, and the second test was carried out to determine the appropriate altitude and length of the parachute and types of containers that would best fit the needs of typical low-altitude drops (no more than 300 feet above ground), keeping in mind that it was necessary for the containers thrown out to clear the plane's tail.

The third test encompassed a real situation that required dropping supplies to Allied units surrounded by the Japanese in the town of Tiddim in the Chin Hills located near the Indian border abutting western Burma. In this case ninety-eight sacked packages that were to be free-dropped (without any parachute or device to slow down the descent), along with eight Mark V and Mark VII containers having 18-foot parachutes were loaded on to a C-46, bringing the total gross cargo weight to 6987 pounds. The total elapsed time for loading the cargo was 45 minutes. To make for maximum effectiveness, three experienced air-crewmen from the 31 Squadron, RAF- were taken on board to supervise the ejection of the cargo with the usual complement of three men in the cargo hold. At an air speed that varied between 115 to 140 mph, the crew of six in the cargo bay ejected the entire cargo with no mishaps or mistakes.

The fourth and final test was a duplication of the third test, only this time the cargo load was increased to 12,000 pounds of both "free" dropped items and containers that were to use 18-foot parachutes. It took the crew 90 minutes to complete the loading, and this time four "ejectors" (also known as kickers) from 31 Squadron were taken on board to facilitate and demonstrate the techniques of pushing supplies out of a plane. Within twenty-six minutes, using eight "ejectors," the entire cargo had been successfully dropped into the designated zone with no difficulty.[61]

Following these tests a number of recommendations were made by TCC: "1:The ejecting crew should consist of eight trained men with "three acting as ejectors wearing harnesses (so as to not follow the cargo out of the aircraft) long enough to enable them to work freely around the door exit." The other five men would pass the load back to the ejectors as they stacked the cargo near the exit making absolutely sure that they were nowhere near the exit; 2:The distribution of the cargo would be carefully monitored by the plane's captain as it was being loaded; 3:An equal number of packages would be unloaded from the front and the back of the plane so as to maintain stability of the aircraft during the drop; and 4:A strongly constructed elevated platform some six feet high with a horizontal top of 6 feet by 4 feet be provided to facilitate loading from trucks to aircraft."[62]

Innovations developed by individual airmen counted significantly in the evolution of air supply and support. Technical Sergeant Charlie Humpal of the 1st Troop Carrier Squadron built a device that would simplify and expedite the dumping of cargo. This device allowed the automatic ejection from the cockpit (containing both pilot and co-pilot) of the cargo with a minimum amount of intervention by the crew.[63]

As these drop runs were being carried out, TCC also conducted tests with C-46s, having them tow two gliders of various weights and for different lengths of time in the air. As noted by Lt. Colonel Willard West in submitting the test results and recommendations to TCC, the C-46 was "excellently suited for troop carrier missions in this theater. They are at present operating from sod strips, taking twice the normal load of C-47 type aircraft. It is considered that they can be operated from any field or strip that the C-47 can be flown from. The C-46 is far superior to the C-47 for glider operations over high terrain. They can tow two fully loaded CG-4A (Waco) gliders over routes and weather that could not be accomplished by C-47s."[64]

To emphasize that assessment, after a series of experiments encompassing C-46s and modifications on the weight bearing capabilities of 22-foot and 24-foot parachutes, TCC concluded that the planes were well-suited to their more demanding roles and that the C-46 could drop as much as 12,000 pounds of cargo as long as the dropping zones were not too small.[65] (Due to the difficulties in maintaining control of the aircraft, the C-46 could not enact difficult maneuvers for drops on tight zones while the C-47 could do so.)

As the training proceeded, the crews encountered other problems. For instance, how might a mule be coaxed onboard a glider, given the reluctance of the animal not to move a muscle when confronted with an unknown situation? How does one package items for air dropping, and what should be used to pack them? Also, in view of the enormous difficulties of climate, and especially the oppressive heat, how far was it possible to push the crews and the machines before one or both would collapse?

Since no manuals existed to direct the loading crews, the men had to load the planes on the basis of common sense. Loadmasters (not having been invented at that point, they would acquire this name later in the war) and air crews, including pilots and co-pilots, plus a number of Chinese co-pilots on the "Hump" run, were directly responsible for the proper loading of their individual planes.[66] As to kicking the items out the cargo door, according to a veteran, one member of the crew would brace himself against the opposite interior wall of the plane from the cargo door with the other two crew members placing the items in front of the kicker. The kicker would then hold on to the fuselage ribs and commence to kick out the cargo when the signal came from the cockpit.[67]

It was assumed that the loading of livestock, including horses, mules, and goats would be a problem. However, the animals, if properly handled by muleskinners and the like, were cooperative, and once they were secured to the interior of the plane the biggest problem was with the animals' excrement. Initially, the use of straw was thought to be the solution to this malodorous problem. However, it was found that the drilling of holes in the belly of the aircraft would facilitate the disposal of liquid wastes, and to some degree the solid waste also found its way into these holes.[68]

At least the animals did not suffer from air sickness, unlike some 60,000 Chinese troops who were flown out of Kunming, China, to Ramgarh, India in late 1943. Almost to a man, every Chinese soldier suffered from acute motion sickness and required a few days of rest to recover from an experience that few Chinese had ever encountered.[69]

As the days approached "Operation THURSDAY's" launch date in the first week of March, TCC and Air Commando 1 had numerous opportunities to hone their skills to a fine degree. In a joint operation of 62 Squadron from TCC and Cochran's Air Commandos, some 192 wounded men were evacuated from the Arakan between February 15th and the 18th. The evacuation was effected with L-5 puddle-jumpers, which were required to land on makeshift airstrips that had been cut out of the jungle close to the front lines. Colonel Cochran followed this effort by graciously requesting the return of his nine L-5s so that his planes and pilots would be properly prepared for Wingate's operation.[70]

As training proceeded and actual combat missions were flown, TCC and Air Commando 1 sustained casualties and damage to their aircraft. In the first eighteen days of February, TCC sustained four aircraft destroyed, and one suffering major damage, bringing the total loss of aircraft in getting supplies to Allied units in northern Burma to six, with one of those lost because the plane had struck a ridge, caught fire, and crashed, killing six of the eight crew members.[71] Unfortunately, training took its toll on men and aircraft, and all too often mistakes resulted in damage and death. TCC and Air Commando 1 were not immune to the vagaries of warfare.

On February 3, 1944, Air Commando 1 was sent on its first official fighter mission in order to orient some of the pilots to the terrain in and around the areas that were designated to be used in Wingate's operation. This operation was followed two days later with a dive-bombing attack on some rail traffic adjacent to the sites of "Broadway" and "Piccadilly," which had been designated as the fields where the Chindits would be inserted, so as to hinder the enemy's response capabilities.[72]

Some ten days later, the first evacuation mission was accomplished, followed shortly by an intensive series of operations that encompassed practically every aspect of Air Commando l's mission in the CBI. By the end of February, Air Commando 1 considered itself prepared for any eventuality during "Operation THURSDAY."[73] Cochran had arrived with specific instructions from General Arnold that, "it was Cochran's job to keep the unit together and do the job that he was sent out for" with no interference from anyone. If that meant having to go to SEAC's commander, Lord Mountbatten, so be it.

Colonel Cochran carried with him a "Dear Dicky" (Mountbatten's nickname) letter that set forth the exact mission of the unit and precisely to whom they would answer. The letter succinctly pointed out that, "When the air commandos get out there, they are to do certain things. They will retain their autonomy. They will not be absorbed into any other commands. They will have one task and one task only, and they are a fastin fast-out outfit. They are not built to last very long. They are going to work for 6 months, and we'll see what happens."[74] The "Dear Dicky" letter was signed by General Henry H. Arnold and countersigned by General George C. Marshall, giving Colonel Cochran unprecedented authority and freedom of action that heretofore had never been granted to a unit of his size in any theater of the war.

With the curtain about to rise on the largest offensive operation so far in Burma, in March, SEAC made its final preparations putting to the test all the experiments, tests, practice runs, and actual combat missions. Whether collectively or singularly, both the British and the Americans had developed methods and tactics to facilitate delivery of supplies to combat units that had been completely detached from their rear areas. To ensure the survival of troops that had been wounded by enemy fire or stricken by the multitude of diseases present in Burma, a novel air ambulance network was needed to assure the survival of the injured soldiers. The unique status of Air Commando One, coupled with the only United States ground unit in all of Asia, GALAHAD, placed these American units in the middle of the largest offensive against the Japanese on land since the war in Asia had commenced. The plan would succeed or fail based on the methods and tactics acquired in the past year in Burma and what the British had managed to put together before the USAAF units had arrived in the CBI. Would all these ingredients, mixed in one exceedingly large pot, provide the means to dislodge the Japanese from their positions in western Burma and push them back all the way to Rangoon, perhaps all the way to Singapore?

## Notes

1) Michael Howard, *History of the Second World War: Grand Strategy, Volume IV* (London: Her Majesty's Stationery Office, 1970), 277.
2) Wesley F.Craven and James L. Cate, *The Army Air Forces in World War II*, Volume IV (Chicago: University of Chicago Press, 1977), 469.
3) Theodore H. White, ed., *The Stilwell Papers* (New York: William Sloane Associates; 1948), 286.
4) Bisheshwar Prasad, *Official History of the Indian Armed Forces in the Second World War, 1939-1945* (Calcutta, India: Sree Saraswaty Press Ltd.; 1958), 12 and according to a memo from TCC to SEAC Hdqtrs. located on roll #A8141, frame #0010, dated 05/13/44, it was determined that all units in the northern and central area of operations in Burma could expect to experience surface temperatures during the day of anywhere from 105 degrees to 115 degrees Fahrenheit and during the hours of darkness temperatures would range from 80 to 90 degrees.
5) Diary, Commanding General, Brigadier General William D. Old, Troop Carrier Command, December, 1983-June, 1944, located at the Office of Air Force History,

Bolling Air Force (AFB) Base, Washington, D. C.; Roll #A8263, frame #001, Title Page, Brief Biographical Sketch. It must be noted here, that the index of Roll #A8263, the listing for this segment is designated as "The General's Diary" with no indication as to which general the diary was of. As of this notation the file will be designated as Old's Diary.
6) Old's Diary, 02/08/44, frame #0007.
7) "History of 330th Airdrome Squadron" located at Bolling AFB, Wash., D. C., Roll #B0726, not dated, frame #250.
8)Ibid.
9) Old's Diary, 02/08/44, frame #0007.
10) Old's Diary, inclusive dates 12/16/43 through 12/23/43, frame #0003.
11) Old's Diary. 02/08/44, frame #0007.
12) Old's Diary, inclusive dates 12/16/43 through 12/23/43, frame #0003.
13) Old's Diary. 02/08/44, frame #0007.
14) Ibid.
15) Interview, Lt. Colonel Philip Cochran conducted by Dr. J. C. Hasdorff, at Bolling AFB, 20, 21 Oct. 1975, and 11 Nov. 1975 at Rochester, NY, p.191.
16) "South East Asia Command War Diaries," "Effect of the Availability of Supplies of SD Parachutes on Operations in 1944," National Archives Repository, Suitland, Maryland. Staff report from Headquarters, South East Asia Command, C-in-C's Planners, Paper No. 7A, 12/29/43," p.1.
17) SEAC Diaries; "Supply Dropping Equipment," Headquarters, South East Asia Command staff report, 12/12/43, p.1.
18) "Ministry of Aircraft Production: Airborne Forces Experimental Establishment, Sherburn-In-Elmet, Near Leeds, Yorkshire," Public Record Office, London, England. Summary of tests on 16 ft. 6 in. parachutes cut on the square, Air 29/516, dated 8.10.1943, pp.1—2. By cutting parachutes on the square rather than the bias which meant that the nylon was cut with the grain rather than against it, the fabric was strengthened to the extent that more weight could be used in the various sizes of parachutes available.
19) SEAC Diaries: "Supply Dropping Equipment," 12/12/43, p.2.
20) Ibid., p.1
21) Ibid., p.2.
22) SEAC Diaries, 01/16/44; Tel. No, 1317 from Sec. of State for India, London, to Gov. General War Dept. New Delhi; Subject: Supply Dropping Equipment.
23) SEAC Diaries; "Supply Dropping Aircraft and Equipment," Headquarters Staff Report, 02/16/44.
24) Ibid.
25) SEAC Diaries: "Parachute Supply Dropping," 12/04/43. p.2.
26) SEAC Diaries, 12/17/43; Telegram on the subject of "Integration of SEA Air Command," p.1.
27) Ibid., p,1.
28) Ibid., p.2.
29) For the most thorough account of the operations of the Chindits refer to Shelford Bidwell's account titled, *The Chindit War,* published by Macmillan Publishing Co., Inc. New York, 1979.

30) The tactical needs of the Chindits called for the immediate response of fighter or light bomber planes to interdict any attempts by the Japanese to interfere with their mission. Major General Davidson, Commander. Tenth Air Force, knew that Air Commando l's mission to be the mobile artillery and striking force for the Chindits was not to be interfered by him or any other commander in the theater. The Chindits had no access to any support on the ground and were absolutely in need of a unit like Cochran's to give them a chance to survive the overwhelming numbers of Japanese that could be called upon to stamp out Wingate's relatively lightly armed units. As General Davidson indicated in an interview some years later in reference to Air Commando 1, "They were totally separate from us. They were with Wingate. When they finished with them, Stratemeyer wanted to give me one or two squadrons, but they promised them they would do this one job and go home, and I knew they wouldn't be any benefit to me, so I wouldn't take them."
31) History, First Air Commando Force, Bolling AFB, Roll #B0680, frame #1404.
32) Interview by phone on May 21, 1993 of Harry A. Blair, enlisted member of 27th Troop Carrier and the Unit's historian, relating personal experiences as his unit arrived in northeast India.
33) "A Historical Report," The Activities of the Twenty-Seventh Troop Carrier Squadron located at Bolling AFB, Roll #A0972, frame #0433.
34) Ibid.
35) "History of the 27th Troop Carrier Squadron" by Harry A. Blair, Historian of the 27th TCS, located in Lacrosse, Wisconsin, p.3.
36) In response to questionnaire mailed out in May, 1993, sent to members of the CBI Veterans Association, Harry A. Blair, enlisted member of 27th Troop Carrier and the unit's historian, relating personal experiences as his unit arrived in northeast India.
37) 27th TCS History, frame #0434.
38) Ibid.
39) SEAC Diaries; Staff Report to C-in-C concerning future operations in the Spring of 1944. "Assessment of Future Operations," 01/08/44, p.2,
40) From Headquarters Army Air Forces India Burma Sector to Commanding General, Army Air Forces, Washington, D. C.;
Subject: Resume of "Thursday" Operation, 03/22/44, roll #A8263.
41) Old's Diary, 01/04/44, frame #0013.
42) Old's Diary, 01/04/44, frame #0013.
43) Old's Diary, 01/14/44, frame #0018.
44) Old's Diary, 01/14/44, frame #0019
45) Old's Diary, 01/16/44, frame #0021.
46) Ibid.
47) Old's Diary, 01/17/44, frame #0022.
48) Ibid.
49) Air Commando l's History, frame #1404.
50) Air Commando l's History, frame #1405.
51) Air Commando l's History, frame #1407.

52) SEAC Diaries, 11/27/43, General Wedemeyer to General Pownall, from Mideast to S.A.C.S.E.A., New Delhi in response to communique #87733 dated 11/26/43.
53) Charles F. Romanus and Riley Sunderland, *U. S. Army in World War II: The China-Burma-India Theater—Stilwell's Command Problems* U.S. (Washington, D. C.: Office of the Chief of Military History, United States Army; 1956),34-35.
54) Chronological History of Tenth Air Force Air Supply Operations 1943-1945, Bolling AFB, Wash., D. C.; not dated, Roll #A8263, p.2.
55) Old's Diary, 01/26/44, frame #0035.
56) Old's Diary, 01/28/44, frame #0040.
57) Old's Diary, 01/31/44, frame #0048.
58) Old's Diary, 02/02/44, frame #0052.
59) Old's Diary, 02/04/44, frame #0053. The British, had by the end of 1943 proceeded with not only the testing of various parachutes, but also they experimented with an assortment of containers that were a necessary contingent of an airborne operation. Specifically, they conducted drop tests of Mark I and Mark III containers that would be suspended with standard 28 foot parachutes, These tests, taking place in August and September of 1943 were specifically designed to "investigate the behaviour of the standard 28 ft. parachute" with a 500 pound load and also to ascertain the maximum speed with which the "container could be launched without serious failure of the parachute."
These tests also included various combinations of materials, including cotton and nylon, which is referred by the British as "Celanese." Included also was a varied assortment of supplies that would in fact be dropped in some future operation, including live ammunition. It was concluded that a 500 lb. container on a 28 ft. parachute falls at a rate of 29 feet per second, whereas a 350 lb. container fell at a rate of 24 feet per second. However, the containers sustained little if any damage and only 1 out of 54 containers sustained any appreciable damage following ejection from an aircraft with an air speed of 125 MPH. The only caveat to that appraisal was that in the event the container landed on one of its edges it would indeed sustain a minimum of damage. In fact, containers purposely overloaded with live ammunition and dropped with either cotton or "Celanese" materials performed almost flawlessly.
It was also determined that once a parachute fitted with a 500 lb. load was used, that the overload of the weight rendered the parachute useless for any repeat usage. However, any weight of 400 lbs. or less enabled the parachute to be used repeatedly. This information was provided by the Ministry of Aircraft Production report entitled, "Airborne Forces Experimental Establishment, Sherburn-In-Elmet, Near Leeds, Yorkshire" dated October 11, 1943. Public Record Office, Air 29/516, p. 15.
60) Old's Diary, 02/08/44, frame #0061.
61) Old's Diary, 02/07/44, frame #0058.
62) "Report on Supply Dropping and Towing of Gliders With C-46" from Headquarters, Troop Carrier Command, 02/26/44, Bolling AFB, Wash., D. C.; Roll #A8263, frame #0264.
63) Report by Hdqtrs, TCC, 02/26/44, frame #0264.

64) Old's Diary, 02/27/44. frame #0100.
65) Report by Hdqtrs, TCC, frame #0268.
66) Old's Diary, 02/13/44, frame #0072.
67) Walter Carre interview. He served as an enlisted man in 7th Bomb Group, India, served in CBI from early 1943 and transferred to China in late 1944. Interview was done at King of Prussia, PA, on May 21-22, 1993.
68) Ralph J, Flint interview. He served as an enlisted man in 3rd Cargo Group based in India from Spring of 1944 with unit transferred to China in I945. Interview was done at King of Prussia, PA, on May 21-22, 1993.
69) William S. Quinn interview. He served as an enlisted man in the 1st Air Cargo Resupply based in India from spring of 1944 thru 1945. Interview was done at King of
Prussia, PA, on May 21-22, 1993.
70) Old's Diary, 02/19/44, frame #0035.
71) Air Commando 1s History, frame #1407.
72) Air Commando 1s History, frame #1409.
73) Ibid.
74) Letter, General H.. H. Arnold, Commander-In-Chief USAAF to Lord Louis Mountbatten, SEAC commander, 12/18/43; Henry H. Arnold Papers, Manuscript Division of the Library of Congress; Washington, D. C.

# CHAPTER THREE

# WINGATE IN BURMA

Major Orde C. Wingate, an unorthodox military figure, played a central role in the China-Burma-India Theater of Operations (CBI). Though an unlikely military strategist, Wingate developed the highly successful strategy which led the Allies to military victory in the CBI. Rather than going through traditional military channels, he built on his personal and political connections within the British government, to swing the Allied military to his ideas. For instance, he influenced Randolph Churchill, son of Prime Minister Winston Churchill, who, in turn, influenced the Prime Minister. With strong political connections – familiar and other – he was able to overcome the obstacles his own military command placed before him. Orde Wingate succeeded in convincing the newly arriving Americans in late 1943, that you could keep troops deep behind enemy lines fully supplied and supported, in continual communication with their base, protected by air cover. In short, he argued you could put troops behind the enemy lines for more than a few days successfully, something that had never been attempted before. This strategy, known as "Operation THURSDAY," was the spearhead for the ground offensive that ultimately pushed the Japanese out of Burma.

As we have seen in Chapter 1, the hardware capabilities of the Allied war effort reached the stage that they could exploited by Wingate and others. Chapter 2 provides the background for a deeper understanding of what the Allied forces were confronting in the CBI. This chapter seeks to connect the discussions surrounding both hardware developments and theories of appropriate deployment in an effort to demonstrate how the Allied command was able to overcome the obstacles and put together a patchwork of a strategy which ultimately proved successful. In this time of crisis, men such as Wingate emphasized the importance of having the right ideas at the right time and place.

In 1940, Major Orde C. Wingate led a successful campaign in Abyssinia (Ethiopia) and was instrumental in returning Haile Selassie back to the throne.[1] His cousin, General Sir Reginald Wingate subsequently penned a letter to his close acquaintance serving in the War Cabinet, General Sir Hastings Ismay. Apparently, General Wingate believed that his cousin's exploits deserved the attention of decision makers back in London. He believed that Orde Wingate's method of warfare, which included an early form of long range penetration would be an important contribution to other theaters of warfare. In fact, he felt confident that Orde Wingate's methods, with certain modifications, could be applied, "to other theaters of operations in our far-flung Empire." [2]

What the senior Wingate was not aware of was that due to Major Wingate's success in restoring Haile Selassie to the throne, he may have inadvertently thwarted British plans in that part of Africa. Regardless of the success in Africa, now Lt. Colonel Orde C. Wingate was not at all well received back in London. There is no question that his attempt at suicide in 1940 while resting and recuperating in Cairo did not reflect well on his state of mind. But perhaps a facet of Wingate's personality also contributed to him being reveled by the upper echelons of the British Army.[3]

With his strong support for the creation of strictly Jewish forces that could be relied upon to fight effectively for the causes of the British Empire, Wingate had by that time firmly embraced the cause of Zionism. This support of an independent Jewish state found little favor with his superiors or with the civilian government. Compounding Wingate's disfavor back in London was his dictating foreign policy by going over the heads of his military superiors and the foreign office with regards to putting Selassie back on the throne. Theses two issues made it more difficult for any possibility of advancement and acceptance of his tactics.[4]

However, Wingate had some influence in the highest levels of government that overrode most of the resistance on the part of his superiors to put his experience to good use. His audaciousness and unorthodox methods had come to the attention of Britain's Prime Minister, Winston Churchill as early as February of 1942. In a letter addressed to Churchill from Professor Harold Laski, Laski suggested the use of Major Orde Wingate in organizing and leading "Chinese guerrillas in Burma, using his Abysinnian technique, to perform Lawrence of Arabia role in Burma." Professor Laski, in another letter to the Prime Minister explained in great detail the lengths that the Imperial General Staff had gone to discriminate Wingate strictly on political grounds. Churchill responded shortly thereafter that, "Major Wingate's talents might be of great value in the Far East Theater," and verbally replied to Laski that "justice will be done."[5]

One week later, Wingate was politely informed by the Military Secretary's department that his current posting in the United Kingdom had been canceled and that he would shortly receive orders to report to General Wavell's staff forthwith. A special messenger was on his way to hand Wingate his orders for the earliest flight to Rangoon. However, a week went by and after having made all the necessary arrangements, he still did not receive his official confirmation. Growing impatient with the delay, Colonel Wingate met with Randolph Churchill. Mr. Churchill suggested that they meet with General Ismay, who might be of some help to move the process along. As might have been expected, Ismay showed no interest in helping Wingate, which left him with one last hope. Randolph Churchill suggested that he put in writing his proposals and he would then pass them on to the prime minister.[6] Apparently, what did move things along was the letter from Wingate's cousin that convinced Ismay that Orde Wingate deserved the opportunity to put his theories into practice.

General Sir Archibald Wavell, the newly appointed British commander of all Commonwealth and Home Forces in India was informed that Wingate would be sent to India, if he wished, and Wavell responded with an enthusiastic, affirmative response and arrangements were made as soon as possible to make the transfer.[7] Lieutenant Colonel Orde C. Wingate arrived in India shortly after the fall of Rangoon in 1942. Having successfully carried out operations in Palestine and Abysinnia (Ethiopia), his exploits had been well-known to his new commander, General Wavell., which was the primary reason for his new assignment in the China-Burma-India Theater of Operations (CBI).[8]

Wingate looked at jungle fighting as a two-edged sword. The jungle could be either an insurmountable obstacle or could be embraced as a natural ally of Allied forces. He had come to realize that the most effective means of damaging the enemy with a minimal amount of men and supplies, was some type of penetration deep inside the enemy's rear areas. The key to success was that a force was needed to be large enough to wreak havoc in the enemy's own backyard, yet small enough to escape once the enemy began to mobilize an overwhelming armed response. Yes, the Japanese were excellent jungle fighters and the jungle represented difficulties in logistics and communication, but he believed there was no reason for British forces not to become as adept at fighting and surviving in the jungle environment as the Japanese had become.[9]

Wingate was the driving force in establishing schools for fighting in the jungle which were established in the summer of 1942, to train British troops in the art of jungle warfare. He also felt strongly that it was unnecessary for the British to be worried about Japanese infiltration, and he was emphatic in the belief that the enemy would be so much chasing their own tails in the process. Insofar as the infiltration of his own units, he had every confidence that, "We had superior brains, resources, techniques, spiritual background, and outlook" to accomplish the task. And, rather than throw away lives for the greater glory of the Emperor as the Japanese had demonstrated so well, his forces would exploit all the above mentioned advantages with as little waste of human life being of paramount importance.[10]

Once Wingate arrived in India, he immediately made an assessment of the situation for the British and introduced what were then considered to be radical ideas concerning the Japanese and the terrain and climate that Allied forces were forced to deal with. His first point was that regardless of the continued success of the Japanese, British forces had no reason to believe their opponent to be invincible. With perhaps a bit of racism and egotism Wingate believed wholeheartedly in the superiority of British forces. He remained unconcerned about future Japanese infiltration. And, while the enemy was using scarce manpower and resources in what he considered to be an enormous waste of time and effort, British forces should in turn make forays deep inside enemy lines with all direct lines of supply and communication cut.[11]

The key to success of any such operation would be efficient and expeditious use of radio communication and air power. Wingate believed that those two components had never been fully exploited and that if used properly, a ground unit could call upon air support for not only supplying a column with food, ammunition, and other necessities, but also for bombing or strafing runs by Allied aircraft enabling the ground unit to escape from any tough situation that the enemy had placed before it.[12]

Implicit in Wingate's theory was the belief that frontal assaults of the opposing armies along some indefensible front-line was not only wasteful of lives and scarce resources, but that the British lines were vulnerable to attack at far too many points that could be adequately defended. Also, why should Allied armies face substantial bloodletting in any sort of frontal assault, when the adroit use of highly-motivated and superbly-led troops attacking the enemy in its rear areas would result in far more damage to the enemy's support network and perhaps undermine the enemy's will to fight? General Auchinleck echoed that opinion in a letter to the Chief of the Imperial General Staff (CIGS) that Wingate, in particular had a value that, "lies in his exceptional power to organize and inspire personally officers and men engaged in highly difficult independent operations and that the further he is removed form personal contact with the troops employed, … the less valuable he is likely to be."[13]

With the use of radio communication, unbroken and ever so reliable combined with sufficient air power, it was theoretically possible to keep a highly-trained and motivated unit in the wilderness almost indefinitely. What Wingate needed above all else in the summer of 1942 was just such a force which he knew would not be available anytime soon. It was not until the beginning of February, 1943, that Wingate had the opportunity to put his theories into operation. Launching his 77th (Long Range Penetration ) Brigade from Imphal, India on February 14th, Wingate commenced a campaign that lasted until the beginning of April. He took some three thousand men deep behind Japanese lines with only radio communication providing a link between his men and the British lines back on the India-Burmese front. Accomplishing little in the way of militarily hurting the Japanese, the 77th returned to India in appalling condition. The unit sustained in excess of eight hundred casualties with the remainder of the Brigade in such ghastly condition, that most had to be hospitalized for several months.[14]

Nevertheless, Wingate was convinced that had his unit not been forced to walk through the jungle to reach their objectives and then return the same way they had arrived, the results would have been significantly more positive. He knew above all else that for a future Long Range Penetration operation to be successful was the ability of a unit to be flown into a designated area, carry out whatever mission they were assigned, and then be flown out the same way they were brought in. To accomplish that task, Wingate needed a large number of aircraft, specifically the workhorse of transport, the McDonnell Douglas C-47 to accomplish the task. However, prior to the arrival of large numbers of United States Army Air Force personnel and equipment, he had to rely on what the RAF had on hand at the time.

In assessing the aircraft that were in use in the CBI between 1942 and 1943, Wingate made some interesting observations. He compared three cargo aircraft, the Lockheed A-28 Hudson, the Douglas Craft DC-2 and its bigger sister, the DC-3. The Hudson could carry a load of 1,400 pounds with a one way range of four hundred miles. The DC-2 could carry a load of 2,000 to 2,500 pounds with a one way range of only one hundred miles more than the Hudson. The Hudson carried five machine guns and the DC-2 was unarmed. The DC-3 on the other hand could carry in excess of 3,000 pounds of supplies and had a flying range, as he put it of being "long." (Its range was listed at 1,600 miles.) Even being unarmed, he found the DC-3 to be his aircraft of choice.

This careful analysis of the aircraft available was also predicated on a hope of his that realistically had little if any chance of coming to fruition.

To Wingate, the Lockheed B-24 Liberator bomber along with its sister long-range bomber, the Boeing B-17 would be ideal for ferrying supplies to his men deep within enemy territory. He concluded that both aircraft could carry at least 6,000 pounds of "stores" with a subsequent increase in the flying range that more than adequately fit into his plans. What is not mentioned is what the reaction of the United States Army Air Forces to the notion of diverting its heavy bombers from the European Theater of Operations (ETO) and the Pacific Theater of Operations (PTO) to carry out what probably was perceived as a decidedly "harebrained" scheme. [15]

Shortly after his return from Burma, Wingate was almost immediately sent back to London at the end of July at the request of Churchill to discuss his plans for future operations. Immediately upon arrival, Wingate was summoned for an interview with the Prime Minister. Churchill was duly impressed with Wingate as a man of considerable imagination and insight, and he took an immediate liking to the man. The Prime Minister was duly impressed with Wingate's plans for a long-range penetration group placed in the enemy's rear by air and decided that Wingate would accompany him, along with the Imperial General Staff, on the *Queen Mary* as it made its way to Quebec for the next Anglo-American conference.[16]

On Saturday, August 7th, 1943 aboard the *Queen Mary*, a meeting was held by the Chiefs of Staff Committee of the War Cabinet. In attendance were General Sir Alan Brooke (Chief of the Imperial General Staff), Air Chief Marshal Sir Charles F. A. Portal (Chief of the Air Staff), Vice-Admiral Lord Louis Mountbatten (Chief of Combined Operations), Admiral of the Fleet Sir Dudley Pound, First Sea Lord and Chief of Naval Staff, and of course, The Right Honorable Winston S. Churchill, M.P. Prime Minister and Minister of Defense serving as Chair. Of the many topics discussed, the primary focus was the future campaign in the China-Burma-India Theater of Operations. They wanted to put into place a plan that not only would re-open a land link with China, but also bring about the re-conquest of Burma with the re-capture of Rangoon of paramount importance.[17]

General Sir Alan Brooke, following a discussion with now Brigadier Orde Wingate, had agreed with the implementation of another Long Range Penetration Group. However, rather than have Wingate's brigade operate in a similar fashion to the first L.R.P.G., this time around the operation had to be backed up by an advance of the main body of ground forces currently situated on the India-Burmese border. Wingate was agreeable to such a plan and added a proposal of flying some 2,000 of his men to Yunnan Province in China to act as a spearhead of the Chinese advance from there. Nevertheless, Brooke cautioned Wingate and subsequently informed the Chief of Staff Committee that such an endeavor would call for an ongoing offensive following the conquering of northern Burma to bring about the eviction of the Japanese entirely from Burma. In addition, Brooke expressed his own doubts as to whether or not that was the best way to go in bringing about the defeat of Japan.[18]

Arriving in Quebec and meeting at the Hotel Frontenanc, the Allies began to put together their plans for the future conduct of the war between the Americans and the British. Along with the Prime Minister, the Imperial General Staff and Wingate came Sir Archibald Wavell, General Sir Claude Auchinleck, President Franklin Delano Roosevelt, General Henry H. Arnold, and the Joint Chiefs of Staff (U.S.A.). It was at this conference that the future command structure of the CBI was established and the green light given to Wingate to go forward with his plans for another operation behind Japanese lines. Field Marshall Wavell was promoted to Viceroy of India, General Auchinleck was appointed as Commander-in-Chief, India and Lord Mountbatten was appointed as the newly created Chief of Combined Operations to be designated as Commander, SouthEast Asia Command (SEAC). It would now be Lord Mountbatten's task to direct all future operations in South Asia.[19]

Immediately following the Quadrant Conference, the Committee concluded that Wingate's plans were plausible and "decided to adopt [a] policy of using them extensively. Based on Wingate's previous operation, the value of such an operation in the future would have a significant impact on overall future offensives against Japan. The nod had been given to Wingate to put into motion all the necessary plans for his second and more greatly expanded operation behind the enemy lines. Wingate came away from the Quadrant Conference firmly convinced that his ideas had finally found a receptive audience.[20]

In conjunction with the creation of SEAC came the proposal to form a special air force with the explicit function of providing all the necessary support that Wingate's columns needed as they made their way deep in the enemy's rear. At the behest of General of the United States Army Air Forces, Henry "Hap" Arnold, the birth of the First Air Commando unit came into being. This unit, with the unprecedented mandate of being the Long Range Penetration Group's private air force, was to be created from scratch with the unique status of limited duration and connected directly with a ground unit that had no United States forces as a component. Arnold chose Lieutenant Colonels Philip Cochran and Johnathan Allison as the men to create this new air force. Cochran, who had gained notoriety in North Africa and Allison, who came by way of Colonel

Chennault's "Flying Tigers" had reputations that indicated that these men were unorthodox in their methods, but achieved results far above more conventional thinking officers.[21]

Having achieved all that he possibly could hope for, Wingate returned to India in September, but before he could put his plans into motion, he was hospitalized in early October with typhoid and was thought unable to be fit for duty at least until the middle of December. As luck would have it, his condition deteriorated to the point that he was transferred to the British Military Hospital in Delhi on the fifteenth of October with his condition listed as seriously ill. A series of telegrams and communiqués between Auchinleck and Winston Churchill ensued, highlighting the concern that the Prime Minister and Auchinleck had for Wingate. Lord Mountbatten also expressed his concern to the point that he made the effort to visit Wingate on a number of occasions. Apparently, Mountbatten was anxious to discuss with Wingate the plans he had in mind concerning the early spring assault east of the Chindwin River, and arranged to have him moved to Viceroyal House as soon as the doctors would allow. As Mountbatten stated, "He can recuperate and attend to a certain amount of business [that] depended upon his personal attention."[22]

Wingate had a number of powerful men in high places that gave him unprecedented authority in the disposition and use of forces unlike any other field commander. That prominence led to the placing of the recently arrived American LRPG (first week of November) under his authority with regard to training. The newly formed 5307th Composite Group (Provisional) was to be placed directly under Lieutenant General Joseph Stilwell for training, but Wingate was able to convince Mountbatten and Stilwell that they should train alongside his own brigades. Whether Wingate was correct on his insistence to have all units slotted for long range penetration operations to train together, it seems only reasonable that a future coordinated assault of disparate units that have undergone joint training could only enhance the prospects for success.

However, even with friends in high places, Wingate was faced with an almost intractable resistance on the part of most senior army officers and most of the high command of the Royal Air Force (RAF). Colonel Derek Tulloch, who served as Wingate's chief of staff was so affected by this hostility, he found it necessary to lay out an entire history and plan for Wingate in the event some documentation was needed to back up Wingate's claims. Tulloch believed that nine out of ten senior army officers would state, if asked, that Wingate's theories were outrageous and impracticable. To refute that consensus, Tulloch insisted that they look at Wingate's career up to his arrival in India and realize that he had more practical combat experience than most any other high ranking officer. Also, the methods that he had developed in Africa and the Middle East had proven uniformly successful and could easily be duplicated in south Asia.[23]

If that was not enough of an obstacle, the planners in Delhi had directed their attention away from any possible major operations in northern Burma using their limited resources on a campaign that would either call for an amphibious assault along the coast of Burma in the Bay of Bengal or a landing on the island of Sumatra in Indonesia. Wingate, on the other hand felt that from a purely psychological point of view, not to take back from the Japanese what was lost by force of arms would have an incalculable negative affect on the morale of all British ground forces. The reopening of the Burma Road was a necessity not only of keeping the Chinese in the war, but also to obviate the need for keeping China provisioned with the only means available at that time, which was the flying in of all supplies over the "Hump," that had put an enormous strain on the pilots and crews who flew the missions. [24]

Adding to Wingate's difficulties was the hostility of the RAF. Perhaps it was more of a turf battle than one of outright resistance, but senior commanders were hostile to any plan that might in theory dilute their control. Given the nature of the jungle environment, with little open space and at most times the inability to clearly identify friend or foe, perhaps they had a point. Nevertheless, Wingate was in no way attempting to usurp their authority or their control. What he had in mind was to duplicate a system of coordination that had been found to be successful in the Middle East. He wanted to centralize the control and allocation of RAF units that could then be put under a "rigorous system in which the airman was given every possible help in the description of the target, the time he was to attack it, how it was to be marked, and various clearly-defined bomb lines which would be seen from the air and beyond which our own troops must not intrude."[25]

Wingate wanted to allocate an initial fifty flight officers to serve as forward observers with his Chindits. Calling on their respective expertise, Wingate believed that these flyers were eminently qualified in choosing appropriate supply-dropping zones and also of landing strips that could accommodate light aircraft for evacuation of casualties. In addition, a radio tuned to RAF frequencies and manned by RAF personnel would engender an enormously comfortable rapport that might never have been achieved otherwise.[26]

Enhancing their resistance further, senior RAF officers contended that there were no portable wireless sets in India that could communicate with planes directly. Only the bulky and non-portable sets used by operations on the ground were capable of maintaining contact. The officers also insisted that since a large number of RAF squadrons had changed from high frequency to very high frequency, there were no portable sets in the entire theater that could be utilized to meet that change.[27]

In refuting that claim, the Brigadier General Staff, Tulloch (B.G.S.) stated that, "The Army 22 set, which we were going to use with our Columns was capable of being used for this purpose effectively." Not accepting Tulloch's statement at face value, it took the intercession of Lord Mountbatten to put the matter to rest. Presiding over this meeting, Mountbatten exclaimed that, "someone here is talking nonsense." After a brief conversation with his signals officer, Mountbatten indicated to the staff that he expected his officer to respond within five minutes for a definitive answer. The phone rang, Lord Louis answered and within thirty seconds slammed the receiver down with a clearly audible bang and turned to Tulloch and said, "You are quite right."[28]

Having settled that dispute, the meeting was adjourned, but not before the assembly was informed that fifty RAF officers were on their way from the United Kingdom for no other purpose than to act as forward ground observers for Wingate's columns. In closing, Mountbatten put to rest any notion that he would tolerate any dissension from his senior officers in the RAF. Mountbatten exclaimed that, "Wingate had been promised certain things, and while I am here I will see that he gets them. Let there be no further argument. His demands must be met immediately for the first three Long Range Penetration Brigades. We will discuss the supply of RAF personnel for the remainder as soon as Wingate himself is available."[29] It is regrettable that Wingate was still hospitalized, for no doubt, this show of support would have accelerated his recovery.

As much as the RAF demonstrated clearly their hostility towards Wingate and his ideas, the same can not be said for how he was received by Lt. Colonel Philip Cochran and his First Air Commandos. The enthusiasm of Cochran came as a most unexpected, yet welcome response to Wingate. Cochran frequently "begged" Wingate and his subordinates to "dream up new methods and ideas for harnessing the air and ground forces in greater unity." Unlike conventional thinking of the time, Cochran believed Air Commando One was not created to play a mere supporting role in the long range penetration groups' operations, but as an active, equal participant in the upcoming offensive. Air Commando One, unlike the RAF, did not need to be asked to provide support. On the contrary, Air Commando One urgently suggested to Wingate what its people could do for the Chindits. Cochran went so far to say that, "You can't give us too much, … we are here to be used flat out, so give us all the work you can." A breath of fresh air to Wingate to say the least.[30]

This unique status of Air Commando One was unprecedented, but there was still much work to be done before any success could be achieved. The Chindits, Air Commando One and the air cargo units coming in from the United States needed to practice and perfect an entirely new type of aerial and ground warfare. Regardless of any prior training and practical experience that the pilots and crews had received up to that point, there were no guarantees that prior knowledge could be effectively utilized in such a unique environment that South Asia presented. The 27th Troop Carrier Squadron had up until their transfer to South Asia, tested and perfected numerous means of supply dropping in and around the Florida Everglades for nearly a year. Since they were not just thrown together quickly or haphazardly, they could in theory, be ready for action upon their arrival in India in January of 1944. Otherwise, all other units of Troop Carrier Command and Combat Cargo Command, along with Air Commando One had no practical experience to call upon.[31]

The first forays by Troop Carrier Command were less than spectacular. Air drops to General Bernard Ferguson in the "Admin Box" in early 1944 were at best haphazard, and at worst, a significant amount of the supplies was landing in enemy hands. Colonel Tulloch, having flown on one such mission determined that the primary problem was that the crew of the aircraft did not have a proper system of pushing the supplies out of the planes. In a memorandum he apprised a number of squadron commanders of the deficiencies and offered suggestions for improvement. Unlike his previous encounters with the RAF, the U.S. commanders welcomed Tulloch's input and unanimously accepted his recommendations. [32]

Assuming that Troop Carrier Command was now firmly on board with Wingate, Colonel Tulloch drafted a memorandum to be distributed to all units in Troop Carrier Command. Outlined in this message was a set of instructions that provided a step by step process of proper dispersal of the supplies that were to be dropped. To Tulloch's surprise, he received a very antagonistic phone call from Brigadier General William Old, TCC commander, asking flat out, if Tulloch had taken command of his unit? As it turned out, Old was totally unaware of the memorandum that Tulloch had sent out two weeks previous. Old then proceeded to search his files to see if indeed Tulloch had provided a copy to him. With all due embarrassment, Old had indeed been sent a copy, but had never been made aware of it by his staff. The General quickly offered an apology for his outburst and they parted in a very amicable manner.[33]

Hurt feelings and egos aside, in early 1944, the strategy of putting a large number of men deep within enemy-held was still more theory than a reality. Wingate had over the previous six months altered his own thinking on the mission of the Long Range Penetration Groups and therefore the scope of the mission had changed. His original conception of the use of LRPGs was of a lightly armed and equipped column inserted deep within enemy-held territory to specifically "raid lines of communication and direct aircraft on to suitable targets, in co-operation with a main offensive."[34]

With the arrival of the United States Army Air Force units, Wingate had to adjust his strategy since it no longer was seen as a shoestring operation. With this departure from his original plans there was no longer practical experience that could be drawn upon to gauge its effectiveness. He now saw an entirely new role for his LRPGs that included not only the deep penetration aspect, but more to the point, his units would become "a spearhead of an offensive in which normal (meaning regular ground units) formations played a secondary role."[35] Only with the inclusion of U.S. units could this be at all remotely possible, but Wingate was thrilled with the changing situation. An opportunity was presenting itself and the ironic part was that he received an overwhelming positive response from his American allies, and continued reluctance and antagonism from his own country's formations.

The Americans, although not particularly well-schooled in the art of pushing supplies out of airplanes and hitting close to their targets in a jungle environment, had much more to contribute to the Chindits' missions than men and airplanes. Specifically, even to the shock and surprise of General Wingate, the Americans had a number of innovations that would prove to be highly effective later on. British headquarters in Delhi was steadfast in its belief that it was impossible to load mules onto a transport plane. According to Delhi, mules could never be forced onto transport planes and if, by chance they were wrong and the mules were actually placed on an aircraft, their behavior inside would have been so atrocious that the pilot would be forced to land.[36]

As the American units were quick to point out, mules and other livestock, if properly handled by muleskinners and the like, could come aboard with little in the way of resistance. As events proceeded in the first three months of 1944, the most serious problem with livestock was with the animals' excrement. As ever innovative in approaching this unusual problem, American engineers drilled holes large enough in the belly of the aircraft (without compromising the aircraft's controls from front to rear) to expedite the elimination of liquid wastes and an unattended consequence was the elimination of a good proportion of the solid wastes.[37]

More significantly, an event in January of 1944 convinced Wingate not only that he was on the right track concerning airborne operations, but more importantly (at least for Wingate's self-confidence) that he had something tangible to show the naysayers in the RAF that they were absolutely wrong about one of his pet theories. Wingate attended a demonstration of a "snatch" of a glider by a C-47 as it flew over the airfield that it was parked on. This technique, which had been successfully attempted back in the United States in 1940, had never been witnessed by any observers of the RAF. By this time, pilots of both Air Commando One and Troop Carrier Command had fine tuned the process to a point in which it was accomplished in a very matter of fact fashion.[38]

Wingate, as enthusiastic as ever, elected to ride in one of the first gliders to be snatched. As the glider's rigging was captured by the slow-moving and low-flying C-47, Wingate was seen sticking his head out of the glider pilot's window and shouting at the top of his lungs, "Go tell the RAF that I have not only seen it done but that I am now doing it too."[39] Providing visible evidence to the RAF that his goals were achievable, Wingate was vindicated by this seemingly impossible feat. He could easily rub the RAF's noses in its shortsightedness and obtuseness, but apparently being proven right was all Wingate needed.

A later demonstration was provided by Air Commando One, only this time the SEAC commander, Lord Mountbatten was in attendance. A British column of some three hundred men and their mules were loaded into gliders and then snatched off the ground by Air Commando One piloted C-47s. Following the spectacle, Wingate turned to Mountbatten and declared that, "This is what I've prayed for," followed by a broad smile on Mountbatten's face."[40] Once again, General Wingate had proved that he was not the madman that so many in the upper echelons of the British Army and the RAF had believed him to be. He had every right to feel good about himself and his unconventional theories pertaining to tactics and strategy in war.

However, even as late as February, only one month before "Operation THURSDAY" was to kick off, Wingate was still denied the appropriate aircraft and trained personnel needed for his plans. In fact, a member of the Imperial General Staff, a General Morris had made his feelings known all too clearly to a party of senior officers by exclaiming that, "Wingate was just a damn nuisance." If that was not bad enough, Colonel Tulloch was continually snubbed by General Headquarters, Delhi in his quest for any new intelligence concerning the enemy. Tulloch had been kept out of the loop so often, that he felt like a "pariah dog."[41]

As disheartening and destructive that may have been for Wingate and his staff, he was continually amazed with the Americans. In practically every facet that included their participation in "Operation THURSDAY," the Americans were only too eager to follow Wingate's lead. The issue of night dropping had been considered impossible by the RAF. Whereas, Air Commando One was willing to attempt it even though they had never been trained for such a delicate operation. Once the unit had been convinced that from the single point of safety, aircraft dropping at night had little to fear from Japanese aircraft or anti-aircraft fire, they became overnight converts to Wingate's proposals. Providing onboard instruction and then a manual that laid out the proper protocols for night dropping, Cochran's men became very adept at hitting their targets, as long as the ground units provided clearly defined drop zones.[42]

In spite of the dead hand of most of British officialdom in India, Brigadier Orde C. Wingate was about to embark on the greatest adventure of his military career. Although "Operation THURSDAY" was to commence at the beginning of March, there was still plenty of preparations to be made and tactics honed to a not too fine a point. His frustrations with the RAF, in particular, would not deter his confidence that his plans would be carried through to a successful conclusion. Obtaining the support of the United States Army Air Force units that General Henry "Hap" Arnold had been so generous in providing, he now had the tools required to do the job. What remained was the underlying doubt as to whether the various pieces of the puzzle could all fit together and make the plan work. Wingate, along with Mountbatten, Auchinleck, Old and Cochran had every reason to believe that the job could be done. But having experienced so many naysayers and antagonists throughout the higher echelons, it was not unreasonable for Wingate to have feelings of self-doubt and persecution. "Operation THURSDAY" would tell the tale.

## Notes

1) Major-General Derek Tulloch, *Wingate in Peace and War* (London: Macdonald and Company, Ltd., 1972), 40-44.
2) Letter from General Reginald Wingate to General Sir Hastings Ismay, Office of War Cabinet 17 January 1942, Imperial War Museum, Department of Documents, London, Wingate Papers, Chindit Box 1.
3) Tulloch's *Wingate*, 47-52.
4) Ibid.
5) Letter from Professor Harold J. Laski to Prime Minister Winston Churchill, 11 February, 1942, Public Record Office, Prime Minister's Office: Operations Papers; Prem3/103/6. It is no doubt very odd that a prominent economist serving in Clement Atlee's shadow cabinet would have written such a letter to the Prime Minister. There is no available evidence in the archives that I had investigated as to why Professor Laski wrote to Churchill about a relatively low ranked officer carrying out unusual maneuvers in Africa and the Middle East. My own speculation is that Laski, a Jew in the British government became very much aware of Orde Wingate's overtly pro-Zionist stance. Perhaps Laski saw in Wingate a man who might make it possible for a permanent Jewish homeland that had been promised by the Balfour Declaration more than 20 years earlier. How many British officers had indicated so prominently not only the reliability of a Jewish Brigade in comparison to a comparable British unit, but also how inadequate any sort of Arab unit would be? Wingate's declaration of the loyalty of the Jews and their capabilities in comparison to anyone else had to have been deemed scandalous by his superiors. Is it any wonder that in a country that had at best ambivalent feelings towards Jews in general, would certainly look upon an officer with pro-Jewish tendencies in its own Army, and do everything possible to keep that officer quiet and out of the limelight?
6) "Narrative of Events," Imperial War Museum, Department of Documents, Wingate Papers, Chindit Box #1, 97/20/6.
7) S. Woodburn Kirby, *The War Against Japan: India's Most Dangerous Hour,* Volume II, (London ,Her Majesty's Stationery Office, 1958), 182-83.
8) Ibid.
9) Tulloch's *Wingate*, 70-71.
10) Ibid.
11) Kirby's *War Against Japan,* 185-186.
12) Tulloch's *Wingate*, 71-72.
13) Letter from General Claude Auchinleck to Chief of the Imperial General Staff, 8 January 1943, Imperial War Museum, Department of Documents, Wingate Papers, CAB 120/805.
14) Bisheshwar Prasad, *Official History of the Indian Armed Forces in the Second World War, 1939-45,* (Calcutta, India, Sree Saraswaty Press Ltd., 1958); 133-137.
15) "Supply Dropping Data" located in the Wingate Papers, Chindit Box 1.
16) Kirby's *War Against Japan*, 398-400.

17) "War Cabinet, Chiefs of Staff Committee: Minutes of Staff Conference" held on Thursday 27, April, 1943 at 3 PM. Located at Public Record Office, London, Prem 3: Prime Minister's Office: Operations Papers; Far East; Prem 3/143/7.
18) Kirby's *War Against Japan*, 380-382.
19) Ibid., 383-384.
20) "Minutes of Staff Conference," 3.
21) "History, First Air Commando Force," Bolling Air Force Base, Washington, D.C., Roll #B0680, frame #1404.
22) Telegram, Lord Louis Mountbatten to General Ismay 2 November 1943, for transmission to the Prime Minister. Located at Wingate Papers, Chindit Box #2, CAB 120/805.
23) "Personal and Confidential Papers" of Colonel Derek Tulloch (hereafter known as Tulloch's Papers).
Located at Wingate Papers, Chindit Box #4, CAB 120/806 (not dated).
24) Telegram, Quadrant to War Cabinet Office, 14 August, 1943. Located at Public Record Office, Prem 3/147/2.
25) Tulloch's Papers, 22.
26) Ibid., 23.
27) Ibid.
28) Ibid., 24.
29) Ibid.
30) Ibid., 25.
31) "History of the 27th Troop Carrier Squadron" by Harry A. Blair, historian of the 27th Troop Carrier Squadron located in Lacrosse, Wisconsin, 3.
32) Tulloch's Papers, 24.
33) Ibid.
34) Ibid., 27.
35) Ibid., 28.
36) Ibid.
37) Diary, Commanding General, General William D. Old hereafter known as Old's Diary. Located at Bolling Air Force Base, Washington, D. C. Roll #A8263, frame #11.
38) Air Commando 0ne' History, frame #1404.
39) Ibid., frame #1406.
40) Ibid., frame #1407.
41) Tulloch's Papers, 32.
42) Ibid., 33.

# CHAPTER FOUR

# THE YANKS HAVE ARRIVED

Burma was only one segment of the British Empire that had been lost by force of arms at the beginning of World War Two in Asia and the Pacific. Because of the threat to India, the need to protect that colony and restore British control in Southeast Asia was of paramount importance. Having lost Malaya in February of 1942, followed closely by the fall of Rangoon, and subsequently nearly all of Burma by May of 1942, the British were now in a completely defensive posture on the Indian-Burmese border, making haste to build some sort of defensive positions against the expected Japanese assault on India itself.

General Joseph Stilwell, having seen his Chinese units nearly evaporate before his eyes after much hard fighting, was left with no alternative but to retreat from Burma as the British had before him. Choosing to make the trek to India on foot, Stilwell and his staff (including 26 Americans, 13 British, 16 Chinese, an assortment of Burmese nurses, some civilians, cooks and mechanics) covered about 140 miles over a fourteen day period. Arriving in India angry at what he and his people had endured, he stated very briefly that "I claim we got a hell of a beating. We got run out of Burma and it is humiliating as hell. I think we ought to find out what caused it, go back and retake it."[1]

Would it be possible for Stilwell to return in kind what his forces and the rest of the Allies had been handed by the victorious Japanese? Given the nature of jungle warfare in a place of such immense proportions as Burma and assuming that the CBI was to be subordinate to other theaters of war, was it remotely possible to resume the offensive in the near future?

The major problems for the Allies were manpower and materiel. These would be exemplified by two separate attempts on the part of the Allies to launch some sort of offensive against the Japanese. Rather than wait for a build-up of all the necessary components for an army on the move, the British launched an attack in the Arakan (just over the southeast border of India) in February of 1943. Almost immediately, the offensive was stopped dead, clearly indicating poor planning and execution. The Japanese were able to employ their favorite tactic of double envelopment, which forced the withdrawal of all British forces before they were surrounded and cut off. Within three months, all British forces had been forced back to their departure point, with little, if anything, to show for their efforts. As General Slim noted in a letter to his superior, Lt. General Irwin, "Whatever happens in Arakan it is imperative that the highest authorities get down to improving the quality of the men in our infantry, and raising their powers of endurance."[2]

Almost simultaneously with the ill-fated invasion of the Arakan, Brigadier General Orde Wingate launched his first Long Range Penetration Group (LRPG) across the Chindwin River from eastern India. With some 3,000 officers and men, Wingate led a campaign against the Japanese rear areas that was meant to show the feasibility of having troops operate in fairly large numbers behind enemy lines, wreaking havoc in the "soft" rear areas. However, as noted by the *Indian Official History*, "The Strategic value of the campaign was nil. It was originally meant to be the spearhead of an Allied offensive but due to a variety of reasons, the expected offensive failed to materialize. The Wingate expedition has been compared to 'an engine without a train.'"[3]

Having sustained some 800 casualties and with the remainder of the men returning in ghastly condition (most were walking skeletons afflicted with all sorts of jungle-borne diseases, making a large number unfit for military duty for a long time), Wingate was still absolutely certain that his LRPG could work. It must be noted that had they been properly supplied throughout their three-month-plus trek through the jungle in both directions, Wingate's Chindits might have accomplished more than blowing up some bits of railway, killing a few hundred of the enemy, and gathering a little useful intelligence.[4]

To enhance the possibility of any sort of success, the British put into motion a number of developments that would be brought to bear in that future offensive. The creation of RAF 31 Squadron on November 9, 1942 was only the first step in a long series of changes that would bring that outcome to fruition. Thirty-one Squadron was specifically designated to provide any and all air transportation and supply-dropping in Burma and the eastern region of India known as Assam. The Squadron would be provided with both C-47s and Hudsons as their primary supply and transport aircraft, and their headquarters would be located in the town of Tezpur. Following the establishment of their base and its requisite facilities, 31 Squadron was to commence with tests and experiments on the dropping of supplies from the air.[5]

Once the unit began operations, developments quickly ensued. It had been concluded that in a number of supply drops, numerous cases of severe damage to planes' tails on the starboard (right) sides had been sustained. Specifically, in supply operations over the Chin Hills (southeastern India on the border with Burma), the cargo being dropped had actually struck the tails of the aircraft, causing so much damage that pilots temporarily lost control of their planes. Also, the parachute rigging lines, having been ensnared by the planes' tails with the cargo still attached, began to cut and strip away the rubber housings of the aircraft's deicers (as was known to 31 Squadron and most likely all units equipped with C-47s, these were Goodrich Deicers).[6]

By June of 1943, Air Headquarters, India, came to some surprising conclusions about the squadron's experiences. Although the C-47 (Dakota) proved to be more flexible and versatile than the Hudson, the aircraft had some problems that if not remedied would be disastrous in the future. The movement of materiel within the cargo area needed to be facilitated in some manner so as to avoid damage to the particular packages and also to prevent the parachutes from becoming ensnared by the plane's tail.[7]

It was enthusiastically suggested that some sort of roller conveyor that would fit the interior dimensions of a C-47 be developed. Within a number of days, drawings of a prototype had been completed. Manufacture of the device was given the highest priority. By the end of June the conveyor was available for implementation.[8]

During these trials, a number of other suggestions and variations were proposed or attempted. It was known that due to the highly unstable air over the Chin Hills, the dropping of supplies below 500 feet was extremely hazardous. Crews were tossed about inside the cargo area as they were attempting to manhandle the cargo out the door, and the accuracy of the drops was severely compromised. Therefore, only during the first five hours of daylight would it be possible to make an accurate drop with little or no injury to the crew. The unstable air currents were lighter during that time.[9]

At about the same time, 31 Squadron also fixed the nature of the cargo crew and the respective jobs of each member. The "throwing-out" crew consisted of three men. Two of the men were to wear belts attached by line to the aircraft. Their job was to "Stack-up" the supplies as near the door as possible. The third man had perhaps the more arduous task of bringing the cargo down the fuselage. He was required to move up to three tons of cargo in anywhere between twenty minutes and one hour, depending on the amount of circuits the aircraft had to make to complete its task.[10]

Assuming the drop zone was approximately 400 yards in length and the loads being dropped weighed 100 pounds each, it was found that a well-trained crew would be able to throw out at the rate of one container per second, necessitating approximately twenty circuits of the target area. If, on the other hand the drop zone was smaller, an additional ten circuits might be needed to complete the job. Additionally, it was found that the stacking of containers near the door was too hazardous to be continued, and that free dropping cargo (without parachutes) was to be effected from heights from as low as fifty feet to as high as 500 feet, whereas parachute drops should be made from a minimum altitude of 200 feet and no higher than 500 feet.[11]

However, as these procedures were being developed, a preliminary investigation was made on a proposal to avoid the use of the aircraft's side door, and instead cut a hole in the floor of the aircraft. This would obviate any concern with parachutes getting caught on the plane's tail and make it easier to drop supplies. On the face of it, it also seemed an easier place for paratroopers to exit the aircraft.[12]

Subsequent investigations concluded that the cutting of a hole was not simply a case of taking a saw and cutting a piece of the aircraft's fuselage away. It was discovered that not only would this cutting undermine the very structure of the plane, but also that the control cables for the flaps would have to be rerouted. Such a drastic design change was not considered practicable within the CBI. The final consideration was that paratroopers preferred jumping from a side door rather than a hole in the floor, because of the lesser frequency of somersaulting when jumps were made.[13]

It was all too apparent that the Allies in early 1943 were unable to effect any large scale movement of ground forces without necessary training and, most significantly, without the proper air support required in the jungle. If the Chindits were going back behind enemy lines for any duration, they needed a secure method of provision supply and the means of calling upon the support of fighters and light bombers to dislodge or destroy stubborn points of resistance. If it was at all possible to obtain such support, given the intensive training and high morale of the Chindits and other special units, it was entirely conceivable that Wingate's theories concerning behind-the-lines operations could bear significant fruit.

At the first Quebec Conference in August of 1943, Brigadier General Orde C. Wingate requested a light plane force to evacuate his wounded men from behind enemy lines in Burma when his Long Range Penetration Group (the Chindits) commenced operations in the spring of 1944. His prior experience in early 1943 with the First Chindit Operation mandated that quick medical attention and evacuation was the only way to enable a large segment of his forces to survive and fight another day.[14]

Accompanying Wingate to the conference was the soon to be designated overall commander of all Allied forces in the CBI, Lord Louis Mountbatten. As Southeast Asia Command's newly appointed commander, his influence at the conference was instrumental in persuading the Americans and the British to provide the tools that Wingate had requested. General Henry H. (Hap) Arnold, Chief of the United States Army Air Forces, was convinced of the necessity to provide unwavering support. Mountbatten's headquarters made tentative plans to form a unit to include light planes and transports for supply operations as well as fighters and light bombers for air-to-ground direct support, in addition to providing escort for the unarmed supply planes.[15]

To head this project, General Arnold chose two fighter pilots with distinguished records: Lieutenant Colonels Philip Cochran and John Alison (as will soon be noted, the choice of Alison proved instrumental in smoothing out some of the red tape). The choice of the two fighter pilots to head this mission was explained, amusingly, by the fact that "they were damn fools enough to try anything that wasn't orthodox."[16] Having made his selections, Arnold also gave them "Carte Blanche orders to gather men and material under the highest authority."[17]

Commencing in late summer of 1943, Cochran and Alison began to comb bases throughout the United States for volunteers. Although many base commanders were reluctant to relinquish their best ground personnel and pilot trainers to these recruiters, with the letter in hand from Arnold, they were powerless to object. In one instance, given prior notice that Cochran's recruiter was on his way, the commander of the field "desired his person and ship elsewhere and that no prospective pilot was to be interviewed."[18] Cochran and Alison, abiding by General Arnold's statement, "to hell with all the paperwork, just get the job done!"[19] aimed to do just that.

Having overcome such obstacles, Cochran and Alison were able to put together a skeleton apparatus and secure equipment, including four of the six experimental helicopters that had just been given over to the United States Army Air Force.[20] As Cochran noted, "We weren't popular to begin with, but when we got there [India], we had some excellent equipment and it was new. It was brought from the states-at least, it was rebuilt."[21]

The organization's designated name (assembled locally), Air Commando 1, code named "Project 9" and then changed to the 5318th Provisional Air Unit, had as its original mission the removal of wounded from the jungle during the operations of the Long Range Penetration Group (LRPG) under the command of Major General Orde C. Wingate. However, this invention of a "new animal in the Air Force zoo"[22] was going to assume a much greater role then originally intended.

This "new animal" was about to embark on a mission that was not to be interfered with, altered, or co-opted by anyone. In a "Dear Dicky" letter from General Arnold given to Cochran to present if needed, Arnold emphatically stated that, "When the Air commandos get out there, they are to do certain things. ... They will retain their autonomy. They will not be absorbed into any other of the commands."[23] This letter gave Cochran unquestionable authority to secure any or all equipment and supplies for his unit and complete autonomy to do as he saw fit in supporting Wingate's operations behind enemy lines.

But this special treatment accorded Air Commando 1 was questionable, for Lord Mountbatten wanted to exert direct control over Cochran's unit. He was preparing to dissolve Air Commando 1 and send its parts throughout the entire region. Cochran was then forced to retaliate with the "Dear Dicky" letter. Mountbatten, after reading the third stipulation in paragraph two, realized that he had little choice but to relinquish control over the unit. In fact, Cochran had the unique authority that only he could communicate directly by cable with the Army Chief of Staff, General George C. Marshall. His cables did not have to go through channels or through Stilwell's headquarters, or any part of Air Force headquarters.[24]

Yet, rather than create poor relations in the future, the tactful Mountbatten said, "You've lifted the whole thing off my back, and I'm so tickled. I can go back with this and tell those guys to hell with it."[25] From that point on, Cochran's unit could perform its mission with no interference. And as it turned out, General Arnold had expected just such meddling from Mountbatten's subordinates and knew that only a threat of direct communications with the Air Force's commander in Washington would keep them at bay. Cochran had his autonomy and access to whatever men and supplies he needed. Now the only question was, would he succeed?

As these events were taking shape, the RAF had previously made some concrete decisions on aircraft they had on hand as early as November of 1942. Through Lend-Lease, the RAF had received more than of 2,000 twin-engine Lockheed Hudsons by 1943. These aircraft, with a maximum air speed of 246 MPH and a range of 1,500 miles, had served adequately as the primary cargo plane for the British.[26]

However, in late 1942, it became apparent at Air Headquarters in Bengal that perhaps the Hudson was not up to the tasks envisioned by the Allies in the CBI. Along with the inexperience of the pilots in the dropping of supplies, the Hudson was deemed unsuitable. The plane was difficult to fly at low speed and when at extremely low altitudes, it was not safe when high rate (tight) turns had to be made; specifically, in the hands of an inexperienced pilot. In attempting to avoid obstructions at the end of a drop run, the aircraft responded sluggishly and on occasion lost some power, resulting in the planes crashing or being seriously damaged.[27]

Perhaps even more significant was the design of the aircraft in relation to its side door and its tail. Apparently, because of the door's location, it was extremely dangerous to load to the top of the door opening. This was in sharp contrast to the design of the C-47, which did not contain these flaws. Therefore, the RAF was limited in its air supply operations until the arrival of additional C-47s in the very near future. (Some 2,000 C-47s would be delivered to all theaters in the hands of the British by early 1945.)[28]

In January of 1944, the first group of men arrived at the bases of Hailakadi and Lalaghat and immediately began training with an airplane strength of: 33 P-51s; 12 B-25s; 13 C-47s; 10 C-64s; 109 L-ls and L-5s; 3 helicopters; and 175 gliders.[29] The officers and men knew that their essential mission was to perform an aerial invasion some 150 miles behind Japanese lines and that they were going to transport troops behind the lines via the gliders that they had brought in with them. In fact, their training included not only snatching gliders off the ground, but also towing two fully loaded gliders with a C-47 without crashing either the gliders or the tow plane.[30]

It was incumbent upon the Allied command, which in August of 1943 was put under one roof known as South East Asia Command (SEAC), to commence a massive buildup of forces in the border regions of India-Burma prior to any future undertakings. In addition, if Lt. General Stilwell was to have any success with his Northern Combat Area Command (NCAC), he would need a large body of Chinese troops. This venture entailed the unprecedented task of removing some 60,000 men from China, with Chiang Kai-Shek's approval, to a training base in northern India, Ramgarh, and commencing a training program under American direction that not only would properly prepare them for fighting, but also provide them with amenities (including receiving silver rupees for their services) that few if any had ever experienced before.[31]

The commander of Allied Forces in the CBI, Lord Louis Mountbatten, summarized the enormous difficulties of an Allied offensive: "An advance in Burma is a very different affair from an advance in France or Russia, since it has largely to be carried out along the single axis of your supply line, and a relatively small force can thus stop the advance of a much larger force, however resolutely led. It thus became of the utmost importance that the overall plans for Burma should prevent large Japanese reinforcements being able to bar the progress of the Chinese-American forces. There were two ways in which the XIV Army could most materially help the advance of the Ledo forces [these were the combined Chinese-American forces of Stilwell's NCAC launching their offensive in the north of Burma from the town of Ledo]. First by cutting the communications of the veteran Japanese 18th Division which was facing the Ledo front, and secondly by engaging the greatest number of other Japanese divisions in Burma. The first task, the cutting of the 18th Division's lines of communications in the south, was given to General Wingate's Long-Range Penetration Forces which included a West African Brigade flown in by Colonel Cochran's U. S. Air Commando 1 force aided by American and British transport squadrons."[32]

If Wingate was going to carry out another long range penetration into the rear areas of the Japanese, he would require the sort of air support never before made available to a unit the size of his Chindits. Wingate deeply disliked paratroopers, but he was very amenable to glider-borne troops. As he wrote in a training manual in mid-1943, "Gliders would avoid stumbling their way through the mailed fist of the enemy which involves them in innumerable hazards before their work has begun and renders them immeasurably less capable of doing it when [the operational area] is reached."[33] And when he was given exclusive operational use of Cochran's Air Commando 1, he likened its firepower to that of a substitute for mobile heavy artillery.[34]

In early 1944, the Allied buildup of men and material was moving along briskly, but the Japanese were not content to wait for the enemy to strike. The Japanese, indeed, were about to initiate an ambitious plan to isolate two of the major crossroads inside India, Imphal and Kohima. These two towns not only served as major staging areas for the Allies, but also as the Allies' only substantial defense positions in eastern India. The Japanese believed that if they could break through these strongholds, the interior of India would be open to their attack.

The Japanese 15th Army was given approval for Operation U-GO on January 7, 1944. The 15th Army commander, Lt. General Renya Mutaguchi was to cross the Burma border into the state of Manipur and seize its capital, Imphal, which was destined to be at the center of SEAC's buildup for the Allies' plans to retake Burma.[35] However, with little artillery and no tanks to oppose the Allies, the only way for this plan to succeed was to perform it with utmost speed.[36] And even though the Japanese managed to amass a good number of aircraft for support, most of them were shot down.[37] Having lost their air support and with little heavy equipment, the 15th Army was left with little except its own tough infantry, relying upon a supply line that extended hundreds of miles back to Rangoon, Burma.

Because of the swiftness of the Japanese assault, a large segment of the British XV Corps under the command of Major-General Frank Messervy had been caught on a large, flat open plain approximately 1200 yards square, surrounded by low wooded hills.[38] Within the confines of this "Box" was not only XV Corps' administrative area, hence the nickname the "Admin Box," but also the supplies built up over the previous three months for the offensive that was to have taken place later in the month.[39]

The question was whether or not the British units, cut off from the rear, would follow the previous tactic of retreating or stay put and kill the Japanese they faced. As observed, the Japanese had severed most of his communication lines with his subordinate commanders, and he and his staff had been pinned down by machine gun fire, unable to make contact with anyone outside of his staff. Messervy, therefore, elected to abandon his position without notifying his divisional commanders, leaving the Box under the command of Brigadier Geoffrey Evans. Evans managed to maintain contact with his scattered units through radio communication and commenced a defense of the Box with no intention of retreating.[40]

The various British units in the Box had substantial firepower to withstand the Japanese. They had many tanks and a large assortment of artillery pieces. Because of the Japanese lack of heavy firepower, and the compactness of the Box, the British were able to react quickly to any Japanese attack. The usual Japanese tactic of surrounding the enemy and demoralizing him, thereby forcing him to withdraw, failed this time; the British did not withdraw.[41]

A great deal of credit for this success must be given to General Old and his Troop Carrier Command. On February 8, Old led a flight of five C-47s into the Box. Old's aircraft was hit eight times by ground fire, which severed one of the elevator control cables.[42] TCC managed to sustain the British effort to the point that SEAC observed, with perhaps a bit of overstatement, that this battle was the turning point of the war. However, Colonel A. J. Barker in his book, *March on Delhi* put it more disparagingly, stating that, "The battle was not of the magnitude which the fuss at Delhi suggested it was. A total of five divisions and an enormous air effort had held off an invasion of one third their strength."[43]

Nevertheless, the Japanese offensive in the Arakan had been halted, enabling the Allies to concentrate their forces on the Japanese further up the more than 700-mile long front that extended from the Arakan in the south to Stilwell's Northern Combat Area Command (NCAC) in the Naga Hills of the northeastern Indian state of Assam.

Although the Japanese offensive towards Imphal had been halted, it was only one part of the overall Japanese plan of overwhelming Allied positions along the Indian-Burmese border. Their most concentrated effort would be made towards Kohima. And ironic though it may seem, SEAC's timetable for a coordinated ground offensive in conjunction with Wingate's "Operation THURSDAY" was scheduled to take place at the same time.

Kohima was not only at a major crossroads of eastern India, but it had also been the pre-war headquarters of the British Deputy Commissioner, Charles Pawsey, which was furnished with a single tennis court adjacent to the commissioner's bungalow. Although General Slim had believed that a Japanese attack would involve no more than a regiment, the Japanese committed an entire division, the 31st, in the first week of April to overwhelm the lightly-garrisoned town (Mountbatten placed the 161st Brigade in Kohima by March 29th). Following the initial assault, the Japanese commenced a two-week siege of Kohima. Having routed British forces from the surrounding ridge lines, they cut the water supply and corralled them "into a triangular area 7000 yards by 900 yards by 1100, and put the garrison under nearly constant artillery fire during daylight hours.[44]

Under such difficult circumstances, the opportunity presented itself for the immediate use of air supply and support to evacuate the wounded from the "Box" and bring in supplies and replacements to keep the Japanese at bay. Initially, because of the size of the "Box," it had been possible to airlift in and out, but as the "Box" became smaller and smaller, not only were planes unable to land, but even parachute drops were falling into Japanese hands. The besieged realized that their only hope for rescue would be by ground units breaking through the Japanese encirclement. In fact, the nearness of the opposing forces had reached the point where the Commissioner's tennis court had become the scene of the most bloody combat where "grenades whizzed back and forth across the width of the court."[45] However, by April 20, the siege of Kohima had been lifted.

Although the Japanese assault could have forced a postponement, Wingate's operation into the Burmese interior, which became officially known as "Operation THURSDAY" was scheduled to commence just before the end of the dry season, sometime in late March or early April. For the Allies to beat the monsoons, they had to commence operations regardless of what was transpiring in another corner of the Indian-Burmese theater. That left Cochran approximately three months to ready his unit. Considering the various obstacles, specifically that their base of operations had been nothing but jungle as late as January 15th, 1944, the Allies' achievements border on the incredible.[46]

A typical day in the first month of operation indicates that their mission was to comprise more than just bringing wounded and sick men out of the jungle and providing tactical support of units in the field. For example, on February 27th, they evacuated 18 patients, put 3 saboteurs behind enemy lines, and dropped rations for men and their mules in 4 different areas. In addition, "Two L-l's in attempting to make a landing on a sand bar on the upper Chindwin, [got] bogged down in the sand and had to be abandoned. The British passengers, however, successfully carried out their intelligence mission and the two pilots walked out nine days later."[47] The following day, T/Sgt. Schatmeyer spied a number of Japanese entering a basha (a small structure made of bamboo, usually used as semi-permanent quarters for officers and enlisted personnel) as he flew his L-5 back to base. As he passed over the basha (which could also be referred to as a barracks), he tossed two grenades out from his aircraft, with one grenade bouncing inside the structure and the other landing directly on top of two Japanese soldiers standing outside, killing them both.[48] Such exploits would soon become routine, all in a day's work.

But aside from the many instances of actual combat with Japanese ground troops, the more mundane need of preparation for "Operation THURSDAY" required a great deal of training to prepare the men on the ground to ready their gliders and planes for the job of ferrying men, mules, equipment, oxen, and supplies to the middle of the jungle. Cochran had realized that it was possible to invade some place behind enemy lines by flying the men in, saving them perhaps as much as a month's time of cutting their way through the jungle. However, having reached the designated area from which they were to operate, it was then imperative to construct a landing strip(s) big enough to accommodate his C-47s while protecting them from Japanese interference until all the troops and supplies were brought in.

Wingate responded to Cochran's queries by insisting that he and his Chindits knew enough about jungle warfare to keep the Japanese at bay and that rather than treat the jungle as just another obstacle, it should be viewed as a friend that could be turned around and used against the Japanese.[49] In fact, what was about to transpire was the invention of "a whole new kind of invasion," and the role that Air Commando 1 was to play was not only going to be greatly expanded, but this new kind of warfare, Air Mobility, was about to undergo unprecedented advancements.[50]

On the 13th of January, the 27th flew its first mission out of Sylhet by flying twelve C-47s to a British base at Argatela, India, where they were loaded up with British supplies that would be delivered to Allied forces fighting for their survival at Tiddim just north of the Arakan. Under the escort of twelve British Hurricane fighters, the 27th commenced parachute drops that called for the C-47s to fly an extremely "tight pattern around a viciously situated objective, where the extremely rough air necessitated accurate judgment and a strong pair of arms and legs."[51] As mentioned previously, the 27th's efforts proved futile, and Tiddim fell to the Japanese. If nothing else, at this particular juncture, the 27th Troop Carrier Squadron (TCS) not only demonstrated the feasibility of air supply and support, but it also proved that given sufficient support in terms of men and supplies, it was quite conceivable that even units encircled by the enemy could survive long enough for either rescue or the accomplishment of their mission.

With the 27th TCS making numerous sorties throughout the CBI and Air Commando 1 putting the finishing touches on its training and active combat organization, when "Operation THURSDAY" commenced on March 5, 1944, they were deemed ready for their most difficult combat trial yet. As will be related in the following chapters, Air Commando l's record of achievement was outstanding. It had delivered on the promise of its special status, which enabled Wingate's Chindits to carry out their respective missions with little if any interference from the Japanese.

However, that is certainly not to say that mishaps did not occur, or for that matter, that all parties concerned were absolutely enthralled by their exploits. As noted by the commander of TCC, General William Old directed some severe criticisms at Cochran's unit. Old was highly dubious about the overloading of gliders, by as much as 750 pounds in some cases, a practice that risked not only the crews on the gliders, but also the flight crews of the tow planes. Although Cochran's intentions were good, Old thought that the risk of the additional weight could have been alleviated had altitude been reduced and course altered, thereby reducing the mileage and strain on the tow planes and all the while reducing fuel consumption and lessening the risk of lost aircraft due to inefficient running.[52]

General Old also criticized Cochran for having failed to coordinate with TCC when Cochran decided unilaterally to add supplemental drops for the Chindits by using C-64s. Old did not deem the using of C-64s a problem as such, but he was concerned because ten of the aircraft had pilots with no previous experience in supply dropping. Thus, three of the aircraft had crashed either upon takeoff or in one instance only fifty miles away from the base. In addition, none of the supplies dropped by the remaining "aircraft were in usable shape when recovered, and the amount of supplies recovered was negligible. This failure threw an added burden of 15,000 pounds, or three additional sorties on TCC as replacement demand was 100 percent."[53]

Taken at face value Cochran's insistence that his pilots were highly experienced in glider towing, General Old allowed Cochran's pilots to take over numerous TCC aircraft as not only pilots, but in some instances co-pilots as well, with the remaining crews made up of TCC personnel. Having witnessed the first day's activities, Old was quick to point out that, "many of Col. Cochran's pilots were not only inadequate as glider tow pilots but also in the fundamentals of formation flying, or even simple air traffic rules."[54]

Unfortunately for Cochran and his unit, General Old, on the day of the operation, arrived at Lalaghat to see Air Commando l's operations firsthand. What he witnessed can only be charitably described as unfortunate: "Colonel Cochran's men were nothing more or less than a mob. No two wore the same uniform, almost all were growing beards. Many appeared idle. Officers and men messed together and shared the same latrines. Sanitation was bad."[55] Old spoke to Cochran of these and other concerns, including the absence of saluting, and Cochran's men did improve operations to some extent, but most of General Old's complaints were either ignored or forgotten due to more pressing issues.[56]

Whether or not General Old's complaints were justified, it is necessary to note that the special nature of Air Commando 1 should have given pause to the clamping down leading to by-the-book behavior and discipline that General Old thought was required for any units directly or indirectly within his jurisdiction. Yes, Air Commando 1 was a sloppy, ill-disciplined unit. And yes, it did bend the rules and cause TCC some discomfort and inconvenience. But, given the nature of its mission and the unorthodox and new methods developed for its special tasks, it would be safe to say that, in principle, a little less conformity and a little more individualism in airmen willing to try either new or unconventional tactics more than compensated for a lack of formal discipline.

As if Cochran had a guardian angel, the SEAC commander, Lord Mountbatten, was pleased with him, noting that Cochran was "an extremely good looking, dashing young leader who has got more decorations for gallantry this war than any other American ace. He was the prototype of the famous "Flip Corkin" the Air Force Colonel in the strip cartoon "Terry and the Pirates" and the speech which he made to young Terry on getting his pilot's wings in this trip was recorded in Congress as expressing most nearly the ideal for which young Americans were fighting. When I told Phil Cochran's men that I achieved the highest honour [sic] open to an Englishman by appearing in the strip cartoon with their Colonel it brought the house down."[57] Also, Lord Mountbatten noted that the men of Air Commando 1 had, "the highest morale of any unit I have seen in this theatre [sic] and it is entirely due to the inspiring personal leadership of Phil Cochran and John Alison and their enthusiastic cooperation with Wingate.

All their men were friendly and eager, laughed and clapped and generally reacted in the most enthusiastic and human way."[58] Is it possible that Mountbatten had seen another Air Commando 1 than the one that General Old had seen? Or was it possible that the SEAC commander's perspective and that of the TCC commander were so different that it was possible for one to see the forest for the trees and the other not?

## Notes

1) Theodore White, ed., *The Stilwell Papers* (New York City: William Sloane Associates, Inc.; 1948), 94-106.
2) Field Marshal the Viscount William Slim, *Defeat Into Victory* (New York City: David McKay Company, Inc.; 1961), 115.
3) Bisheshwar Prasad, ed., *The Official History of the Indian Armed Forces in the Second World War: The Reconquest of Burma*, Vol. 1 (Calcutta, India: Sree Saraswaty Press Ltd.; 1958), 135-136.
4) Louis Allen, *Burma: The Longest War, 1941-1945* (New York City; St. Martin's Press; 1984), 146-47.
5) From Air Headquarters, Bengal to 31 Squadron Detachment, 1/09/42. Public Record Office, Air 23/2035, London, England.
6) 31 Squadron, R. A. F. Kharagapur to Air Headquarters, Bengal, 05/28/43. PRO, Air 23/2021, p. 1.
7) Ibid.
8) Air Headquarters, India to Air Headquarters, Bengal, Subject: Equipment and Paratroop Dropping--Dakota Aircraft, 06/17/43. PRO, Air 23/2021.

9) 31 Squadron to AH, Bengal, 1.
10) Ibid.
11) Ibid.
12) Ibid., 2.
13) Ibid.
14) Michael Howard, *Grand Strategy, Volume IV* (London, Her Majesty's Stationery Office: London; 1970), 545-550.
15) History, First Air Commando Force, Bolling AFB, Roll #B0680, frame #1404.
16) Ibid.
17) Ibid
18) Ibid.
19) Ibid.
20) Interview, Lt. Colonel Philip Cochran conducted by Dr. J. C. Hasdorff, at Bolling AFB, 20, 21 Oct. 1973, and 11 Nov. 1975 at Rochester, NY, p. 189. It must also be noted that Lt. Colonel John Alison was a close friend of Harry Hopkins. Originally, the first group of helicopters was to be given over to the British. Both Cochran and Alison saw the potential of the helicopter for their unique mission and by using Alison's influence with Hopkins, they were able to convince Hopkins to intercede on their behalf and get first shot at the helicopters.
21) Ibid., p. 188.
22) Air Commando l's History, frame #1405.
23) Cochran interview, p. 190.
24) Ibid.., pp. 196-197.
25) Ibid..
26) Enzo Angelucci, *The Rand McNally Encyclopedia of Military Aircraft* (New York City: The Military Press; 1980), 313 and 325.
27) Ibid.
28) Ibid.
29) Air Commando l's History, frame #1406..
30) Air Commando l's History, frame #1407.
31) Air Commando l's History, frame #1409.
32) Ibid., 193; 195; 199; 202..
33) Ibid., 179.
34) Ibid., 178.
35) Ibid., 179.
36) Ibid., 183..
37) Ibid., 184.
38) Ibid., 188-89.
39) *The Longest War*, 228-35.
40) Ibid.
41) Ibid., 236-38.
42) Ibid.
43) Air Commando l's History, frame #1409.
44) Ibid., frame #1407.
45) Ibid.
46) Ibid., frame #0433.

47) Ibid.
48) In response to questionnaire mailed out in May, 1993 sent to members of the CBI Veterans Association, Harry A. Blair, enlisted member of 27th Troop Carrier and the unit's historian, relating personal experiences as his unit arrived in northeast India.
49) 27th TCS History, frame #0434.
50) Diary, Commanding General, Brigadier General William D. Old, Troop Carrier Command, December, 1943-June,1944; Bolling AFB, Roll #A8263, frame #0117.
51) Ibid.
52) Ibid.
53) Ibid., frame #0118.
54) Ibid.
55) Ibid.
56) Ibid.
57) SEAC Diaries, April 9, 1944, "Extracts from Supreme Commander's Diary," National Archives Depository, Suitland, Maryland, 1.
58) Ibid

# CHAPTER FIVE

# IT'S SHOWTIME

"Operation THURSDAY" the code name for Major General Orde Wingate's Second Expedition behind the Japanese lines in Burma, commenced on March 5, 1944. Not only would Wingate's Chindits be ferried by glider to a base of operations some 150 miles behind the front lines, but in conjunction with this massive air assault, Lt. General Joseph Stilwell and his Chinese forces in the Northern Combat Area Command (NCAC) would launch a ground assault against the key railroad hub of Myitkyina in northern Burma. Wingate's forces were placed not to meet heavy Japanese forces head on, but to interdict the supply lines of the Japanese 18th and 56th divisions that would ultimately face Stilwell's NCAC forces.[1] What made this entire operation unique was that Wingate had at his disposal Air Commando 1 under the command of Lt. Colonel Philip Cochran. Unlike any other air unit created in the United States Army Air Forces, Cochran's command was established to serve but one master for a specific duration and then be allowed to go home.[2]

As indicated in SEAC's notes on the Wingate operation, General Henry "Hap" Arnold, following the appointment of Lord Louis Mountbatten as SEAC commander, had offered any assistance required. Mountbatten then indicated to Arnold that Wingate's forces, once established behind enemy lines in Burma, lacked any way of being extricated from their positions and, therefore, needed some sort of "air unit which could establish two-way radio communications for the LRPGs operating behind the Japanese lines, to enable supplies to be flown in and casualties, etc. to be rapidly evacuated."[3] As of that moment Air Commando 1 was born.

For the operation to achieve initial success and to prevent the Japanese from guessing Allied intentions, the three landing areas designated PICADILLY, BROADWAY, and CHOWRINGHEE were to be "kept clear of all Allied aircraft in order not to draw the enemy's attention to them."[4] Also, the respective distances of these landing sites in relation to the bases in India from which "Operation THURSDAY" would be launched gave credence, from the Japanese perspective, that it was of no more importance than other points in northern Burma. BROADWAY was situated some 268 air miles from Hailakandi, 267 miles from Lalaghat, and 182 miles from Imphal. PICADILLY was 269 miles from Hailakandi, 268 miles from Lalaghat, and 184 miles from Imphal. And CHOWRINGHEE was 250 miles from Imphal.[5]Hailakandi, 248 miles from Lalaghat, and 172 miles from Imphal.[6]

With such an unprecedented deep penetration into enemy territory, the operation was intended to concentrate Wingate's forces in the area around the town of Katha, which was located at map coordinates "96 degrees 20 minutes East and 24 degrees 10 minutes North, cutting the enemy rail and river communications with his forces opposing the Chinese forces advancing down the Hukawng Valley, and those operating against the IV Corps on the Chindwin."[7] The units chosen for the mission were the 77 Brigade, which would be moved to PICADILLY and CHOWRINGHEE, taking off from fields at Hailakandi and Lalaghat, and two columns of the 111 Brigade, which would be moved from the same strips, with the remaining six columns leaving from Imphal. All units were to be landed at PICADILLY and CHOWRINGHEE. BROADWAY was designated as an alternate landing site.[8]

In particular, 77 Brigade, was a logical candidate for this unprecedented assault. As recently as the previous January, the Brigade had ironed out a number of difficulties that were to be expected during "THURSDAY" operations. One of the most important problems that needed to be addressed was the coordination of the units on the ground and the aircraft providing vital coverage for them. In particular, the liaison between Army and Air officers had to be one of mutual understanding and clear purpose for all parties involved. Wingate believed very strongly that the only way to maintain an efficient relationship was to have the RAF liaison officer be as experienced on the ground as he was in the air. As Wingate put it, "The difference to a Column Commander between having one of our R. A. F. fighting officers to consult on air problems, or having the usual semi-educated liaison officer, is the difference between confidence and uncertainty, between efficiency and incompetence, between vitality and emasculation."[9] There is no question that proper coordination of supply drops and air to ground fire support would be instrumental for the success of the operation.

In addition, Wingate had been able to eliminate doubts about of having supplies dropped in the dead of night. It had been accepted fact that pilots would have enormous difficulties in finding their proper drop zones under the cover of darkness. However, following a series of supply-dropping experiments near Manipur from January 19 through the 21st, an entirely different conclusion was reached. Not only did the pilots have no more difficulties in finding their way at night, but they actually preferred the freedom of having no fighter escorts to coordinate with, and that flying at night was easier on them both physically psychologically.[10]

TCC was assigned as director of all transport operations, including providing twenty-six pilots as co-pilots and thirteen transports as tugs for the glider operations assigned to the 5318th Air Unit (Prov). The C-47s designated as tugs were originally required to carry some "1,000 pounds of cargo to be dropped simultaneously with the release of the tow rope in the dropping area adjacent to each landing strip. Due to the difficulties in obtaining altitude with the double tow of overloaded gliders, it was decided to remove the 1,000 pounds in each tug."[11] The materials that would have been carried by the tugs were considered essential for the troops immediately upon landing, so rather than risk the safety of the tugs and the gliders, TCC was asked to provide six additional C-47s to carry the cargo. This change most likely would have been the best solution, but the 5318th Air Unit (Prov) decided, without consulting TCC, to load the remaining cargo into ten of its C-64s to be free dropped over BROADWAY. As it turned out, the crews of the C-64s had no training in supply dropping. The cargo that managed to reach the target was unusable and the 5318th lost three C-64s in the process.[12]

Prior to the implementation of the air invasion of northern Burma, Air Commando 1 had "carried out training in night dropping of supplies from C-47s to perfect night dropping techniques and to effect coordination with ground personnel."[13] Flying out some 20 to 25 miles from their base, Air Commando 1 flew these trial runs over brush country that was generally level, but contained numerous hills, buttes, and other natural obstacles. This difficulty required Air Commando 1 to coordinate with the ground units various methods of identification using flares, blinkers, smoke, and brush fires quickly to pinpoint Allied positions. Other training included day and night takeoffs with Waco gliders either in single or double tow, pick-ups of gliders both loaded and unloaded from the air (snatches), and night snatches of gliders. Also, Air Commando 1 pilots carried out reconnaissance and familiarization flights into Burma where they experienced combat conditions and suffered some combat losses.[14]

TCC included thirty-nine C-47s from the 27th Troop Carrier Squadron (TCS), the 315th TCS and a TCS from the 5318th Air Unit, and forty-four C-47s from the 31, 62, 117, and 194 Transport Squadrons from the RAF.[15] Special facilities were then constructed at all three airfields to facilitate the loading of troops, animals, and equipment on to the planes. Loading pens on the runways were marked from 1 to 40. All items to be loaded were divided into 6,000 pound loads and assigned pens: "When an aircraft taxiied [sic] into a pen, loading began. Bamboo stalls were built in each plane for transporting animals. They were constructed so that when the animals were not being transported, the stalls would not interfere with other cargo. Ramps for loading animals were provided for each loading pen. Servicing of the transport went on concurrently with the loading of the aircraft."[16]

Overall, SEAC anticipated that it would take seven weeks from planning to actual implementation of the plans to carry out "Operation THURSDAY."[17] The day of the operation was set for March 5th to take advantage of a full moon, and of course "Operation THURSDAY," like any major effort in the CBI, had to commence before the onset of the monsoon season.

As noted by Mountbatten, the hardships that the LRPGs would encounter behind enemy lines not only included the enormous difficulties of terrain and climate, but also the fact that following the creation of SEAC, the Japanese had increased their Burma forces from four to nine divisions. This addition brought the total number of enemy troops deployed in Burma to some 250,000 men, creating the highest concentration of Japanese forces anywhere in the world outside of China.[18] He further indicated that, "Unlike other theatres [sic] where troops can be moved and supplied by sea, in the interior of Burma troops can only be successfully moves [sic] and supplied by air, unless we are prepared to spend the next 20 years building roads behind every advance and then having our lines of communication cut by the Japanese."[19] Therefore, Mountbatten concluded, as did General Arnold, that the uniqueness of Burma called for never-before-attempted tactics that would negate the problems of terrain, enemy dispositions, weather, and lines of supply and communication. "Operation THURSDAY" was to be just such an attempt, with no guaranteed outcome.

One of the unforeseen circumstances that confronted the Allies just one hour before the commencement of "THURSDAY" was the shocking discovery from some last minute air photos taken of PICADILLY. The landing area of PICADILLY had been "completely blocked by the dragging of large trees, in lines across the field making it definitely unsuitable for glider operations."[20] With only fifteen minutes left before the first wave of C-47s and their towed gliders were due to take off, a hasty conference of all air and ground commanders was convened to ascertain the situation. Because a British C-47 had used PICADILLY as a pick-up point in the spring of 1943 to evacuate some of Wingate's men from the first LRPG and had been popularized in various popular magazines, it was concluded that the Japanese had placed the logs there on purpose.[21] (As it was later discovered, dozens of teak logs at PICADILLY had in actuality been put there by Burmese natives as a way to dry the logs out in the open before the wood could be processed.)

Should the entire air assault be postponed? Had the Japanese somehow become aware of the Allies' plans? And what about BROADWAY and CHOWRINGHEE; were they also compromised, and if so, what should be done? It was pointed out that even substituting CHOWRINGHEE for PICADILLY was not possible since the air crews had not been properly briefed for a full-scale glider operation on that strip.[22] Also, "it was believed that to call the operation off when the troops were loading in the gliders would have serious repercussions on morale."[23]

Following consultation with all parties, including Lt. General George Stratemeyer; Lt. General William Slim, Wingate's operational commander and XIV Army commander; Air Marshall Sir John Baldwin, commander Third Tactical Air Force; Brig. General William Old; Major General Orde Wingate; and Lt. Colonel Philip Cochran,[24] the decision was made to proceed with the air assault with the substitution of BROADWAY for PICADILLY. This change was made "possible since the arrangements made for landing reception and parking of gliders at BROADWAY and PICADILLY were identical, and therefore crews who had previously been briefed for PICADILLY would be equally at home if switched to BROADWAY."[25] However, because of the change of landing site, and due to the assault taking place under the cover of darkness, the original number of 80 gliders was reduced to 52 gliders to avoid congestion at BROADWAY and avoid accidents on the crowded field that BROADWAY would ultimately become.[26]

One last problem encountered by the Allies was the difficulty of the double tow. Although the towing of two gliders by C-47s had been practiced successfully within the previous six weeks or so, the release of the tows on the day of the assault resulted in a number of gliders being prematurely released, and in a number of instances the tows had parted enroute. It was, therefore, decided that "later departures were to be all single tows" and the "Single tows were, moreover not so hard on DAKOTA engines." To make up for the number of gliders that had not reached their appointed destination, the decision was made to send an additional fifteen gliders in single tow to BROADWAY, but on arrival they had to be waved off due to congestion at the landing site; fortunately, all managed to return to base safely.[27]

Ironically, because of the losses in the initial assault, the crashes of gliders all over northern Burma caused considerable confusion for the Japanese.[28] Because Standard Operating Procedure (SOP) in combat required the turning off of all aircraft running lights once they had reached the Imphal Valley, when some of the tow ropes broke, the glider pilots had no way of knowing where they were and immediately lost sight of their tows. Ten gliders, having been left to fend for themselves, made emergency landings at Tulihal (which was in Allied hands), along the east bank of the Chindwin River only a few thousand yards from the Japanese headquarters at Pinlebu. In one amusing incident, a glider carrying a Ghurka detachment landed at Tulihal in the mistaken belief it had landed at BROADWAY. Following a good landing by the pilot, the Ghurkas began streaming out of the glider before it came to a complete stop with guns at the ready to shoot at anything moving in the darkness. Fortunately there was no one within the immediate vicinity of the aircraft, and the Ghurkas were very shortly advised as to their whereabouts.[29]

Air Commando 1 pilots Major Taylor and Lt. Colonel Alison were the first glider pilots to be towed over the landing site at BROADWAY. Its gliders not only carried assault troops, but "field fighting equipment and radios." Having landed without incident, the troops immediately set up a defensive perimeter and arranged the landing lights for the landings that were to follow immediately (the U.S. glider crews, unlike their British counterparts, were not designated to join the units that they brought in, but to wait patiently for a 'pick-up" back to their bases). It was soon discovered that there were several teak logs hidden from view in the tall grass and that tracks from logs that had been dragged on and off the field had also been obscured by the overgrowth of tall grass. Because of this unseen obstacle, numerous gliders had their undercarriages ripped apart, making it exceedingly difficult to move them out of the way of later arriving gliders. A number of gliders crashed into the gliders that had been disabled on the landing and two gliders fell into the jungle just a few yards short of the landing strip. And although there was some radio communication between units on the ground and gliders that were still in the air, one of the gliders carrying vital radio equipment failed to arrive.[30]

The confusion and damage sustained on the first assault wave, however, did not deter or substantially slow down the landings at BROADWAY. Colonel Cochran, upon learning of the confusion in the first wave, ordered all gliders to be towed as singles once the tow planes returned from their first run.[31] Fifteen gliders that had reached BROADWAY safely had to be returned to base since the congestion at the landing site had not been cleared up. Overall, of the sixty-seven gliders which had successfully taken off from their bases in India, only thirty-two managed to land at BROADWAY. Of the remaining gliders, nine landed in enemy-held territory and eleven landed or crashed in areas in Allied hands.[32]

Increasing the confusion of the Japanese, six of the gliders that had been lost were located by pilots from Cochran's force: "They were scattered from a point just east of the Chindwin, right in the middle of the Jap lines to a point about 40 miles from BROADWAY and it was a foregone conclusion that altho [sic] the probable loss of personnel was tragic, the tactical advantage that must have resulted from the enemy being confused and bewildered at the idea of striking forces at ten different places in his rear was tremendous and gave impetus to the operation's success."[33]

Regardless of these positive aspects of the invasion, the situation at BROADWAY had indeed become acute: "The inability to clear the landing strip was in large measure due to the fact that the gliders carrying the mechanical equipment had failed to reach their destination and, therefore, all work was restricted to manual effort."[34] Because of the loss of PICADILLY, and with enemy reaction to the invasion still unknown, a substantial part of the glider assault had to be canceled. The unavailability of the heavy equipment (e.g., bulldozers) necessary to compensate for the mess at BROADWAY seriously hampered operations. Twenty-three men had been killed, and twenty-three had been injured in glider crashes up to that point. Nevertheless, over 400 men had arrived safely along with a considerable amount of supplies.[35]

Of immediate concern to the Allies was the necessity of getting BROADWAY prepared for the landings of C-47s. Conflicting reports of the situation on the ground indicated that the "fly-in" plan of the cargo planes had to be postponed until late afternoon of March 6. In addition, of all the gliders that had made it safely to BROADWAY, "only three were serviceable enough to be towed and snatched out."[36] As these events were unfolding at BROADWAY, twelve gliders had been single-towed to CHOWRINGHEE. Two of the gliders that crashed at that landing site unfortunately contained the bulldozers and other heavy mechanical equipment needed to clear the field for the landings of the Dakotas, thereby leaving CHOWRINGHEE also without heavy equipment.[37]

Following the frenzy of activity at BROADWAY and CHOWRINGHEE, the Allies could expect an immediate Japanese response to this deep incursion behind their lines. No reaction seemed to be in the offing, but to ensure that the Japanese would not interfere with the assault, Air Commando 1 with its P-51s and B-25s followed Japanese movements south of the invasion area, primarily on the Japanese bases at Shwebo, Onbauk, and Anasikan. Air Commando 1 managed to destroy forty-five Japanese aircraft on the ground, with a loss of only two P-51s.[38]

Also, to confuse the Japanese further, Air Commando 1 had two P-51s "with cables hanging from the bomb racks swept out to break wire and telephone communications between vital Jap points." And to intensify the confusion, "The Strategic and Tactical Air Forces cooperated in the mission by bombing nearby Jap airfields, dropping dummy paratroopers and making numerous tactical sorties."[39]

By the morning of March 8th, the Japanese still seemed to be unaware of the events at BROADWAY and CHOWRINGHEE The landing area at CHOWRINGHEE had been made suitable for C-47s, and a total of eighty-four C-47s and five gliders had flown into both landing fields on the night of March 7th, through the early morning hours of the following day. Those additional flights brought the strength of the Allied forces to about 3,000 men, with their equipment and a large number of mules to assist in moving the heavier items. And as yet, there was still no significant response from the Japanese.[40]

Given the negligible Japanese interference, the Allies stepped up their operations considerably. Some two thousand feet of makeshift runway had been readied at BROADWAY, which set in motion the bringing in of the entire Long Range Penetration Group with all of its equipment. At the airdromes of Lalaghat and Hailakandi, RAF and TCC planes were packed with men, mules, and supplies and readied for their sorties into the combat zones. The operation was a complete success except for one casualty, a specially designed bulldozer that was needed to clear away heavy objects. Without it, the preparations of CHOWRINGHEE were significantly hampered, at least initially. (Regrettably, the three crewmen on the glider were killed in the crash.) Having received the radioed message on the loss of the equipment, the commanding officer of the 5318th Air Unit (Prov) immediately dispatched a C-47 to Calcutta for a replacement. At 1930 hours, the bulldozer arrived at Lalaghat and was immediately transferred to the glider scheduled for takeoff to CHOWRINGHEE at 2100 hours. A short time later the glider came in safely, and the needed heavy work at CHOWRINGHEE was accomplished.[41]

Within twenty-four hours, the situation at both BROADWAY and CHOWRINGHEE had become as normal as at any busy airport in the United States. "Hordes of transports were coming into BROADWAY with split second timing. Col. Alison had set up his field in a manner that would make the controller at LaGuardia turn green with envy. Ships were scheduled two minutes apart and were coming in right behind each other, unloading and roaring back into the air. There was a tremendous dust hazard but the pilots proved themselves equal to the job and carried thru [sic] with perfect timing." The situation had been so drastically improved at BROADWAY that by the second night (March7/8) BROADWAY was able to accept a landing every forty seconds.[42]

As Colonel Cochran related after the mission, it was crucial to preserve the secrecy of the operation. Therefore, to prevent the likelihood of the initial landing party being overwhelmed by superior forces on the ground, "the rule was, that once the places were picked, photographed, studied, and decided upon, we would not keep flying over there."[43] However, as noted before, the landing strip at PICADILLY had been blocked by teak logs placed there by the Burmese, and the decision was made to bring the gliders into BROADWAY and CHOWRINGHEE and forego the use of PICADILLY.

Once that decision had been made, the actual landings were not incident-free. Although the glider pilots had been trained to guide themselves after release from their tows by following the exhaust flares from the tow plane's engines which gave off a clearly discernible blue light, this procedure proved to be a hazard that at the time was completely unforeseen. Not only were the glider pilots forced to guide themselves by a few blue dots in the sky, but they also were being "low-towed," obviating an opportunity to orient themselves with the ground while handling the equivalent of overloaded, unwieldy flying boxcars. To compound the glider pilots' difficulties, Air Commando 1 on numerous occasions exceeded the limit of 4,000 pounds specified by the British. Cochran noted that, "We practised [sic] this, and we knew every piece of equipment, and every person, and everything that was going into each glider. Each glider was assigned its load and its personnel. They were brought up to the zone, and right up the loading area in excellent, meticulous order. And we had practised it over, and over, and over again. We actually practised everyone of those complements that were going in with their equipment and they would stand back, and then on signal, they would trot forward and learn how to get in that airplane and strap themselves in and go. We had that down to a science."[44] Despite having prepared themselves for all foreseeable circumstances, mistakes were made and not all the men and equipment reached their designated landing zones intact.

However, mishaps did not in any way deter the Allies from proceeding with "Operation THURSDAY." To dramatize the nearly perfect operation, General Old took it upon himself to lead a flight of C-47s into BROADWAY on March 7th. Having transferred the loads of twenty-seven gliders into the Dakotas, General Old led this flight to BROADWAY and found the situation rather amazing. Colonel Cochran had done such an excellent job in air traffic control that the average time on the ground for a C-47 was about twenty minutes. Only one major incident occurred that day. Two Dakotas had collided while taxiing for takeoff, causing ten injuries to crew and passengers and forcing the abandonment of the aircraft for the time being. On General Old's flight in, his complement of troops had barely disembarked before his plane filled up with the crews of the gliders that had either crashed or been disabled. Of all the gliders that had landed at BROADWAY, only three were serviceable enough to fly.[45]

By noon on March 8th, TCC had fallen some sixty sorties behind schedule. In all, some 2,200 men had been delivered to BROADWAY with their equipment, including 360 animals along with equipment needed by Air Commando 1 on the ground, and 180 Gurkhas had been delivered to CHOWRINGHEE along with the machinery needed to clear the landing field. Within the previous twenty-four hours, ninety-one transports and four gliders had been flown in to BROADWAY.[46]

As the landings were taking place, Air Commando 1 was not going to wait for the Japanese to attack from the air to disrupt the landings at CHOWRINGHEE and BROADWAY. Japanese air strips at Indaw and Katha were systematically bombed and strafed by P-51s and B-25s. And yet, there had been no Japanese response by the morning of the 8th. Puzzling as the situation may have been, a flight of P-51s was directed to reconnoiter Japanese airfields in the Anisakan-Shwebo area. Over Anisakan approximately seventeen Japanese fighters were found on the ground. The fighters dropped some forty-two 500-pound bombs on the runways and revetments and followed this sortie up with numerous strafing runs on the aircraft parked around the field. According to the pilots' after-action reports, every single Japanese aircraft was left either totally destroyed or burning to the ground. Lieutenant Krug was credited with the burning of five Oscars on the field.[47]

Following the attack on Anisakan, the returning planes flew over the enemy airfields at Shwebo and Onbauk where over sixty fighters, bombers, transports, and two training planes were found. As Air Commando l's planes made their first strafing and bombing runs, they found the airfield to have the appearance of an almost peacetime condition with planes, fuel, and supply trucks scattered haphazardly throughout the fields. The fighters and bombers began their attacks. After exhausting their ammunition, they tallied some twenty-seven Oscars destroyed on the ground and one destroyed in the air; six Dinahs, one Sally, and one transport destroyed on the ground; numerous buildings were left in flames; and three gasoline trucks were left engulfed in flames. Because there was more work to be done, a message was sent back to base, and a flight of B-25s was immediately dispatched armed with fragmentary and incendiary bombs to finish the job.[48]

The B-25s patterned-bombed the airfields, leaving twelve aircraft on fire and numerous other fires. Overall, a total of forty-eight aircraft were confirmed destroyed, which according to Allied estimates comprised twenty-five percent of the known Japanese aircraft strength in all of Burma.[49] Why the Japanese were caught off guard has never been clearly ascertained, but a great deal of credit must be given to careful planning by the Allies. Also, the Japanese were certainly spread very thinly within their interior lines of supply and communication.

The fly-ins of C-47s continued on the nights of March 9, 10, and 11. Ninety-five C-47 sorties were put into BROADWAY, and forty sorties were put into CHOWRINGHEE on the 9th. This activity completed the phase of the operation for CHOWRINGHEE, and the following night the field was abandoned as portions of Wingate's LRPG had made their way into the jungle to begin operations to interdict any Japanese response to Stilwell's forces attacking from the north. BROADWAY, on the other hand was maintained as a major airdrome and built up by more sorties on the night of March 10/11. The final four sorties were flown in the following evening.[50]

Overall, "Operation THURSDAY" comprised 579 C-47 sorties and 74 glider sorties that transported 9,052 men, 1,359 mules, 254 tons of equipment, one Bofors anti-aircraft battery, and one 25 pounder artillery piece. All flights had taken place under the cover of darkness over a period of six consecutive nights. Not a single transport was lost and only six C-47s sustained minor damage from the landings themselves.[51]

Following the abandonment of CHOWRINGHEE, Japanese fighters attacked the makeshift airfield sometime after 0800 hours on the 10th of March and again in the afternoon. What they found were the burned out remains of eight gliders that had previously been damaged in the landings and were too far gone to be of any use. All personnel and equipment had been removed the previous night, including six gliders, which were snatched out. The situation at BROADWAY remained quiet until March 13th, when two waves each of fifteen Japanese fighters attacked the field. Having received warning of the enemy's approach, six Spitfires were able to get airborne in time to shoot down four enemy aircraft. One Spitfire was lost.[52]

Cochran's Air Commandos had been observing all Japanese activity originating from the Japanese airfields at Schwebo and Onbaak. On the morning of the 9th, 13 P-51s were loaded with two 1,000-pound bombs each and ordered to bomb the runways and storage areas at the Onbaak field. Encountering no opposition in the air, the P-51s commenced their bombing and strafing runs, destroying one Sally and damaging one Oscar on the ground. Subsequent reconnaissance revealed that both enemy airfields looked to have been "burned out," negating the need for any further attention to those enemy installations.[53]

The enormous success of "Operation THURSDAY" cannot be minimized. Not only had the fly-ins to BROADWAY and CHOWRINGHEE not been seriously interfered with by the Japanese, but the remarkable success of flying in thousands of men and mules and their equipment at night over a six day period with so few mishaps and casualties was totally unprecedented in the history of modern warfare. The Japanese had been almost totally ignorant of the events transpiring in their own backyard, and when they had finally made a show of force, it was easily defeated. By the 12th of March, air operations against Japanese airdromes resulted in the loss of sixty-three aircraft destroyed and at least eighteen damaged (one attack against the Japanese airdrome at Heho resulted in the destruction of 17 aircraft and the damaging of 11 others on the field, with the loss of only one P-38). When the Japanese did attack BROADWAY and CHOWRINGHEE, enemy losses were heavy, whereas Allied aircraft lost were minimal.[54]

A C-47 had been sitting at BROADWAY for four days with a damaged aileron (the control mechanism located in the tail) and no replacement part available. It would have ordinarily been destroyed by Allied forces rather than be left to fall into enemy hands. Having flown a C-47 only once before, and more than a year ago at that, Lt. Colonel Alison flew the plane out of BROADWAY as it was. Having made it back to Lalaghat with little difficulty, he then managed to get his landing gear down, stabilize his aircraft, and land the plane safely.[55]

General Old did, however, indicate his frustration with the British and in particular Wingate's inflexibility about the loading of transport aircraft. According to General Old, the practice of loading aircraft was directed by Wingate's staff, and in accordance with its doctrine, the loading of planes followed one and only one procedure, with no deviation whatsoever. Specifically, the British did not consider proper weight distribution of the cargo inside the aircraft. As General Old put it very bluntly, they paid no attention to gravity problems in a/c [aircraft] loading, the loads had been to the tail, and the officers so instructed stuck by the book almost to the end."[56] Apparently, TCC and the RAF disagreed on the proper method of combat loading of transport aircraft, but their lack of rapport did not hinder the effectiveness of air supply and support for the Allied units in the field.

There is no question that the feeble response of the Japanese to the Allied invasion enhanced the success of "Operation THURSDAY." General Wingate informed General Old on March 11 that the Japanese commanding general in Burma had informed Tokyo that the Allied operation was "only a diversion by 1400 troops in 80 gliders" and that "the information had come from a soldier captured in one of the gliders which broke its tow rope."[57] The success of "Operation THURSDAY" was due to good luck with the weather (on only one night out of the six were there clouds, threatening a heavy downpour), and Japanese arrogance certainly did not hurt the effort.

Had this operation taken place in an area heavily occupied by a well-supplied enemy equipped with radar, the story might have had a radically different ending. However, it was proved beyond a doubt that air mobility played a part in the offensive against the Japanese in Burma. Furthermore, the amount of cooperation between disparate air forces and ground units demonstrated that such operations were possible.

With the first phase of "Operation THURSDAY" completed, the Allies could view their accomplishments with considerable pride and confidence. Some 9,000 combat troops, 1/2 million pounds of weapons and ammunition, and 1,300 animals had been flown into the heart of northern Burma without the loss of a single man or aircraft to enemy action.[58] In addition, the cooperation between the USAAF and the RAF was unprecedented. Even though each organization had its own grating idiosyncrasies and methods, as Allied units they managed to put those obstacles aside so as to achieve the objective.

As Wingate's forces moved out from BROADWAY and CHOWRINGHEE, he sent a message to Mountbatten to be relayed to Prime Minister Winston Churchill. He extolled the accomplishments of his forces to protect himself from his critics at SEAC and London. The general indicated that CHOWRINGHEE had served its purpose well and had according to plan been abandoned once his columns were moved out. And at BROADWAY, not only had his 77 Brigade blocked all surface communications at Mawlu, but other columns had advanced to their objectives without notice by the Japanese, while at the same time maintaining BROADWAY as their "Stronghold." The BROADWAY stronghold, in fact, was to serve primarily as a light plane airport (meaning the liaison aircraft of Air Commando 1) bringing in necessary reinforcements and supplies and the removal of casualties.[59] General Wingate had every right to feel vindicated about his theories of a quite unconventional means of striking at the enemy.

Having made their way into the interior, Wingate's forces, although highly trained and motivated, still could not expect to carry out their missions on their own and return to India reasonably intact. Air Commando 1 had carried out the first part of its assignment in nearly flawless fashion. Once the lightly armed columns had gone their separate ways with little firepower, in the event of a strong Japanese response, the Chindits would most certainly need assistance that was available only from the air. To ensure clear communication and understanding of the needs of the ground forces, each column had attached to it an RAF observer serving as ground controller. Having trained before "Operation THURSDAY" in India, Air Commando 1 and these ground controllers had developed tactics and methods that would best use the firepower available from Cochran's air force. All the RAF observers had been completely familiarized with American jargon and had met many of the pilots back in India. To enhance the ability of Air Commando 1 to hit targets in the densest of jungle, a number of B-25s had made prior photo reconnaissance of the most likely areas that Wingate's forces would be operating in once they made their way from BROADWAY and CHOWRINGHEE.[60]

The process of this coordination would be as follows: Since all the pilots and the ground controllers involved had access to these photos, the chances of mistaken coordinates were minimized; mortar smoke marked the target; attacking a target was only permitted after confirmation from the ground controller, thereby lessening the chance of friendly fire errors and waste of the time and fuel of Air Commando 1. Such techniques required time and a great deal of skill and clear communication among all parties concerned. It took a little while for the ground troops to feel comfortable with friendly aircraft operating so close to their own positions. However, fighter pilots hit their targets with a high degree of accuracy. Many columns had Cochran's P-51s hitting targets within fifty yards of their own positions.[61]

By March 18th, 77 Brigade, under the command of Brigadier Calvert, had set up an all-important roadblock at Mawlu to intercept any rail or road traffic heading north towards Stilwell's forces approaching Myitkyina. This roadblock, which would acquire the nickname "White City" because of all the parachutes that had been dropped there to supply the Allied units, was where "the majority of the fighter and bomber activities were directed."[62] Also, four days later, another area known as "Aberdeen" was carved out of the jungle with a landing strip that could accommodate C-46s.[63] In essence, "White City" and "Aberdeen" were a replay of BROADWAY and CHOWRINGHEE, indicating clearly that the units on the ground were able to move in the almost impenetrable jungle terrain, leave one base hacked out of the jungle, and move on to any other place and be able to establish themselves with little if any interruption in their missions or their supply.

By the 18th of March some 2,200 men and 1,000 mules had been put into key positions around the important juncture of the town of Mongmit. Not having to plow their way through dense jungle and bypassing the hazardous crossings of both the Chindwin and Irrawaddy Rivers enabled 77 Brigade to dig in and prepare a nasty surprise for the Japanese. Not only had air transport provided the means to bring them to their positions, but the continued support by air drops kept them properly provisioned and lacking in nothing to proceed in their mission of interdicting the Japanese.[64]

However, to throw the Japanese into more confusion, 77 Brigade devised an air drop to convince the enemy of one intention, whereas in reality the Brigade's intentions were directed elsewhere. The Brigade wanted to convince the Japanese that their targets were the towns of Goktiek, Maymo, and ultimately a blocking of the important Mandalay-Lashio road. It was hoped that the Japanese would redirect their efforts in that direction, especially if Allied actions pointed to a threat to their headquarters in Maymo. To convince the Japanese of this ruse, a number of clear messages unencrypted were exchanged to lend credence to the presence of more Allied units than there actually were.[65]

However, this tactic did not mean that what Wingate's forces left behind was entirely abandoned. Although CHOWRINGHEE had indeed been left for the jungle to reclaim, BROADWAY was still in operation. In one instance, because BROADWAY was still occupied, a flight of nine B-25s and eight P-51s was able to use the landing strip there, after being unable to penetrate a solid weather front moving in over the Chindwin that effectively blocked the route home to India. The pilots and their planes spent an uneventful night in Burma, and at daylight the next day flew back to India. Because BROADWAY was still active, another party of downed glider pilots and crews, having eluded Japanese patrols for ten days, found their way to BROADWAY and were immediately flown back to India for well-earned rest and recuperation.[66]

All in all, the month of March had been highly fruitful for Air Commando 1. The total number of sorties flown by Cochran's fighters was 424, including 161 bombers, 80 gliders, and 516 transports. As the Allies had made their penetrations deep inside Japanese territory, Air Commando

1 took an enormous toll on the enemy's forces on the ground and in the air. Some fifty aircraft were destroyed. Also, two river steamers were damaged, and twenty-nine motor transport were destroyed. In and around the rail yards, Air Commando 1 destroyed forty-eight boxcars and tanker cars. They pulverized four locomotives, and with eleven more damaged, knocked out eight bridges, and thirty-eight warehouses were demolished. These targets of opportunity, known as soft targets, also included seven ammunition dumps destroyed along with miscellaneous buildings and barracks.[67] With such a varied and vast scorecard of destruction and damage done to the enemy, Air Commando 1 had proved its effectiveness not only as a supply unit to the ground forces, but also as a swift moving mobile flying artillery and gunnery unit that could bring quick response to targets pinpointed by the RAF controllers as well as other targets of opportunity.

But these were the totals only for March. As the units that had set up "Aberdeen" and "White City" were about to find out, the Japanese did finally become aware of the dangers that were lurking deep within their own lines: "White City," held by Calvert's 77 Brigade, situated north of the town of Mawlu astride the rail line connecting Mandalay with Myitkyina and approximately halfway between BROADWAY and CHOWRINGHEE held by General Calvert's 77 Brigade. "Aberdeen" located some twenty miles northwest of "White City" was held by Brigade Major Master's 111 Brigade. Calvert's unit, having cut the railway, followed that with the destruction of three major bridges. The entire defensive perimeter was "a rabbit warren of slit trenches and burrows roofed over with logs proof against anything smaller than a direct hit from a heavy artillery shell. The perimeter wire was deepened into an obstacle sown with booby traps and further protected by outlying minefields, and area immediately in front of their position was covered by the interlocking fire of machine-guns and mortars centrally controlled through an intricate web of buried telephone cables radiating from Calvert's command post."[68]

"White City" by the 4th of April began to take the shape of more than just a temporary block in the middle of the jungle. Five gliders made their way to a makeshift strip loaded with engineers and heavy equipment. Having all the necessary earth moving equipment, the engineers in less than twenty-four hours built a strip able to accommodate C-47s. By the night of April 5/6, C-47s began routine landing operations into "White City." Cochran made note of some comments made by a captured Japanese soldier who had been keeping a diary that chronicled the soldier's experiences since he was transferred to the Burma theater. The soldier remarked that "The enemy transports were landing and taking off all during the night and the lights on their wings making the Japanese uncomfortably unconcerned." Cochran followed this observation by his own, stating that these operations had by now become relatively routine and their accomplishments an accepted fact.[69]

Having made preparations for a defense against a Japanese assault, and having been amply supplied by the arrival of C-47s, Calvert's brigade now only needed to wait patiently for the Japanese. Starting on the night of April 6th, a large body of Japanese, under the command of Major-General Hayashi, launched an assault after a three-hour bombardment. As the fighting continued through the night, Calvert called for an air strike at dawn and Cochran's Air Commandos responded with an attack by six Mustangs that forced the Japanese to withdraw for a time. As the fighting continued through the day, liaison aircraft were able to land at the makeshift airstrip located between the railway embankment and the jungle covered-block. They ferried the wounded to BROADWAY, where they were loaded onto a C-47 bound for India.[70]

By April 7th, the Japanese intensified their assaults against Calvert's positions and brought up more artillery; on the 9th they deployed two small tanks. However, amply supplied by air drops of mortar bombs, grenades, and machine gun belts, while limiting the amount of food rations due to the demand for ammunition, Calvert's forces managed to beat back the assaults by the Japanese. Some 700,000 belts of machine-gun ammunition were parachuted to the 77 Brigade in "White City."[71]

As part of its continuous support for all the Allied forces at "Aberdeen" and "White City," Air Commando 1 kept a relentless pressure on movements by the Japanese. In particular, after adding five new P-5lBs (a newer model than the P-51As that arrived in India with him) to his armada, Cochran immediately put these longer-range fighters to work. On April 21st, the fighters attacked the Spa Schweli bridge. This mission required a round trip of 750 miles, with each fighter loaded with two 1,000 pound bombs. The Mustangs, using dive bomber techniques that Air Commando 1 had perfected, scored direct hits on the suspension bridge's center span and tower, thereby dropping the entire span into the river.[72]

Also, on the 21st of April, Air Commando 1 began having L-5 observers use mortar smoke bombs to mark targets that were not clearly visible to faster moving fighters and bombers. The L-5 pilot would fly at tree-top level, drop his smoke bomb on a target, call in the attack aircraft, and hurry out in time to see the enemy position pulverized by the accurate bombing and strafing of the P-51s and B-25s. For the month of April, Air Commando 1's score was as follows: 719 fighter sorties; 186 bomber sorties; 350 transports; and 19 gliders. They destroyed thirty-five Japanese aircraft, four motor transports were demolished, plus twenty-one damaged. Three bridges were destroyed and six units of railroad rolling stock had been left in ruins. And at least forty-five buildings of various types were left burning.[73]

To further indicate the skills acquired by Air Commando 1 pilots, as the fighting became more intense at "White City," it became necessary for Calvert's men to call for fighter support at extremely close range. Following the marking of targets by mortar smoke from ground units, an attack was called down on a Japanese force massing for a counterattack less than two hundred yards from the British forces. Armed with 200 and 500 pound bombs, twenty-seven Mustangs came in and deposited their deadly cargoes right on top of the Japanese positions. The sheer volume of high explosives literally shook the earth and following the accurate direction of ground fighter directors, rendered the entire battlefield quiet and still.[74] Once again, the skill of the P-51 pilots and the pinpoint accuracy of their targeting enabled the Allies to forestall, if not prevent a massive enemy assault. These combined arms operations resulted in the 5,500 men of the Japanese around "White City" sustaining more than 3,000 casualties.[75]

The events of March and April 1944 had proved that proper training, coordination between disparate forces on the ground and in the air, and highly motivated personnel, a new type of mobility for ground troops could be achieved. Not only could large numbers of heavily equipped men and their machines be airlifted to targets deep inside enemy-held territory, but once there, they could be sustained nearly indefinitely, provided the air arm had the skills and determination needed to protect those units.

To summarize the experiences of the ground units up to this point, Flight Lieutenant R. G. K. Thompson submitted a report as an appendix to the "Report on Operations of 77th Indian Infantry Brigade in Burma" provided by General Wingate prior to his untimely and tragic death on March 24th (the C-47 carrying him and his staff crashed into a mountain in the Chin Hills; yet it clearly had no impact on the success of the operation). Thompson's observations indicate quite clearly some of the difficulties that had to be overcome in preparation for another operation in the near future. He noted in particular that the war being waged against the Japanese did not fall within the parameters of contemporary warfare as it was being experienced in other theaters of operations. Specifically, he observed that "There are no great targets the destruction of which will prove fatal to the enemy, at least not until we are in a position to attack Japan herself. Instead there are a multitude of tiny targets of small importance individually and mostly either unrecognisable [sic] from the air or easily simulated. The enemy lives widely and methodically dispersed."[76] Therefore, Thompson concluded that only an observer in close quarters on the ground with a LRPG unit could effectively bring in a response from the air that would prove effective in neutralizing that position.[77]

As far as Lt. Thompson was concerned, the use of pack animals in the future had been rendered obsolete. In his opinion, "Pack transport-mules, elephants and coolies is limited by the number of men and animals available, the inadequacy of their load and the slowness of the movement through such difficult country. For tactical reasons long penetration groups must not be encumbered by a pack train that hinders their speed and mobility, and that prevents them from tackling natural obstacles that they might otherwise be able to cross."[78]

Therefore, Thompson concluded that future LRPGs must eschew their animals and coolies and rely on air transport entirely for the columns' needs throughout the operation. As radical as that may seem, he pointed out that 77 Brigade relied on three Hudsons and three Dakotas for their needs during "Operation THURSDAY." Given the nature of the size of a typical LRPG unit, 2,200 to 3,000 men, he believed that as long as there were enough supply aircraft available on a moment's notice, the ground unit could manage very nicely.[79]

Also, for the first time in World War Two anywhere, the first functional helicopter (capable of carrying passengers), the VR4 (built by Sikorsky), was put into service. In early May, a VR4 was used to evacuate wounded men who could not be moved over land, from positions that were completely inaccessible to ordinary aircraft. The helicopters were able to make their pickups with no untoward incidents and transport their precious cargoes to nearby airfields for quick evacuation by C-47 to India. The only problems with these new rotary planes encountered by Air Commando 1 were mechanical difficulties that developed due to the newness of these aircraft and the enormous toll taken on them by the tropical climate.[80]

As with the helicopters, continuous air operations over nearly three months and the climate exacted a heavy cost on the men and machines of Air Commando 1. Numerous cases of dysentery, malaria, and exhaustion had begun to influence Air Commando 1's effectiveness. Many of the pilots and ground personnel were hospitalized. As May arrived, so did the monsoons, which curtailed operations significantly. However, the Japanese were not going to wait out the monsoons; they kept the pressure on the Chindits as the latter made their way north to continue their task of blocking Japanese attempts to send reinforcements against Stilwell. And even though Air Commando 1 was supposed to end operations with the Chindits by the first of May, they were still called upon to provide much-needed support for the LRPGs. As the British were in the process of evacuating "White City," along with "Aberdeen" and BROADWAY, they requested that another airstrip be placed astride their line of march along the railroad. In doing this, the British planned to bypass the Japanese stronghold at Nopin and place their roadblock above them at Pinbaw. A glider strip was subsequently constructed, enabling supplies to be ferried in, but the Japanese put up some effective ground fire. They destroyed one glider and damaged a C-47 as it was making its landing.[81]

With the approach of the monsoon, Colonels Cochran and Alison were "sent back to Uncle Sugar [USA] to form cadres for other Commando Groups" and the other officers and enlisted personnel were transferred to a new base at Asansol, India, where they were able to recuperate, rest, and train for future operations.[82] An assessment of Air Commando l's campaign is needed at this point, covering the previous three and one half months of activity that includes operations before, during, and after "Operation THURSDAY." As noted in the unit's "Official History," Air Commando 1 made a clear distinction between the operations of its fighters and its bombers. Of the 253 P-51s deployed throughout the period, of which 198 were in commission, only nine fighters were lost, in the air or on the ground. This figure encompassed a total of 230 missions flown which amounted to a total of 1,482 sorties. The total amount of bombs dropped on the enemy was 1,156,530 pounds, including high explosives, incendiaries, depth charges, and frag clusters, even though 358 of the sorties were without any externally mounted ordinance.[83]

Air Commando l's B-25 bomber force was, however, a significantly smaller unit. Only 11.3 bombers were officially assigned to the unit, an average of only 10.2 in commission at any one time. Of that total, five of the aircraft were lost on the ground or in the air during operations. Yet, this small force in which each plane was capable of carrying up to 4,000 pounds of bombs, flew 182 missions, totaling 422 sorties that delivered 431,167 pounds of bombs on the Japanese and their installations.[84]

And just as important as delivering deadly cargoes on the Japanese, Air Commando l's force of 180 L-ls and L-5s flew 7,257 combat missions resulting in the evacuation of 1,865 wounded men, and the dropping or flying in of rations enough to feed 9,700 men and 400 mules. This hectic schedule resulted in an average of forty-seven combat missions for every pilot.[85]

The enormous impact of Air Commando 1 cannot be judged purely on the events leading up to and including "Operation THURSDAY" but also must take into account their on-call status for providing close-in support to the Chindits. As indicated by one unit in the field, the Third Indian Division's intelligence summary indicates quite clearly the immense contribution that Cochran's force made to its efforts on the ground. A report dated March 24 stated, "Air action in support of our ground troops was reported to have been literally deadly accurate," and that "shortly after the air support, the enemy retreated from this area having previously established LMG [light machine guns] and sniper posts."

Other comments made throughout the rest of March and April include the following: "March 26: 40 enemy killed and 6 warehouses destroyed as a result of air action. March 29: 460 of enemy were killed in action at Indaw. Effective close support A/C did yeoman work." April 15: "This third battle of the Nenu block has seen larger forces disposed by both sides than other previous attacks. Furthermore, air action on this occasion against the enemy has been consistent and destructive. Amongst other things it has been shown that aircraft alone can force the enemy to move or leave its artillery." And April 19th: "Due to excellent collaboration between ground and air, targets were bombed and shot up within a very short distance of the perimeter of our defences [sic], Our troops on the ground described the results as excellent."[86]

Cochran's men and machines performed beyond anyone's expectations and erased any doubt in the minds of British and American commanders that given the proper tools and more importantly the proper communication between the air and ground forces, a small force of highly trained and motivated pilots equipped with perhaps the finest propeller driven aircraft of its time could perform feats that as recently as one year previously could not have been envisioned. Why else were Cochran and Alison sent back to the "States" to organize other Air Commando units with the blessing of the Commanding General of the United States Army Air Forces, General Henry "Hap" Arnold, if not to take advantage of the methods and doctrines they developed in combat?

Colonel Cochran's Air Commando 1 and General Old's Troop Carrier Command had provided the means necessary to make General Orde Wingate's theories on operations behind enemy lines a reality. Although Wingate had been killed in March in the mountain range west of Imphal, the success of "Operation THURSDAY" transcended the loss of that unique individual.[87]

Throughout Burma in April and May of 1944, the pressure put on the Japanese on all fronts was bearing significant fruit. With the enemy using some 100,000 of its best troops in south Asia, locked in what would be a disastrous assault on Allied lines on the west bank of the Chindwin, facing Lt. General William Slim's XIV Army, plus the assault from the north of Lt. General Joseph Stilwell's Chinese divisions toward Myitkyina, the Japanese had been stretched beyond their logistical limits.

Moreover, the Japanese were also under assault from deep within their own lines by Wingate's Chindits. The commanding general of the Burma Area Army, Masakazu Kawabe, did not have a single reserve formation capable of sustained combat available to repel the intrusion of the Chindits in his own backyard.[88] These multiple responsibilities for the Japanese in Burma made operations that much easier for the Allies, but nevertheless the air invasion required enormous amounts of skill, bravery, materiel, and dedication.

In consequence, the combination of an ill-fated offensive carried out by the Japanese at Imphal in which 65,000 of its own 155,000 men perished,[89] plus the Allied offensive in northern Burma and behind Japanese lines, turned the Burma campaign completely around in favor of the Allies. By May, it had become apparent that it was possible to strike at the enemy's enormously vulnerable rear areas with almost total impunity as long as all facets of a military force could coordinate their movements and demands. The signal job of Air Commando 1 had proven that a heretofore risky and exceedingly dangerous mission could be carried out, as long as all the human factors could be subordinated to the common goal. Of Air Commando 1, it may be truly be said: as they were needed, they were there.

## Notes

1) Shelford Bidwell, *The Chindit War* (New York City: Macmillan Publishing; 1979), 78-9.
2) History, First Air Commando Force, Bolling AFB, Roll #B0680, frame #1407
3) SEAC Diaries, 02/05/44, Major General Playfair to Commander-In-Chief, 11 Army Group, Subject: LRPG Operations 3rd Indian Division, 2.
4) Report on Operation "THURSDAY" from Air Commander, Third Tactical Air Force, Air Marshal Baldwin to SEAC, EAC, XIV Army, Third Indian Division, TCC, Air Commando 1, Strategic Air Force dated March 17, 1944, Bolling AFB, roll #A8143, frame #612.
5) Report of Troop Carrier Command Participation in "THURSDAY Operation" from Headquarters, Troop Carrier Command to Commanding General, Eastern Air Command dated 03/16/44, Boiling AFB, roll #B0726, frame #0248.
6) Ibid.
7) Ibid.

8) "Report on Operations of 77JH Indian Infantry Brigade in Burma-February to June 1944" by Brigadier 0. C. Wingate. London, England, Public Record Office, Air 23/1943, p.8.
9) Ibid., 24.
10) Report of Troop Carrier Command Participation, frame #0249.
11) Ibid.
12) From Headquarters, Army Air Forces, India Burma Sector to Commanding General, Army Air Forces, Wash., D.C.; Subject: Resume of "THURSDAY" Operation, 03/22/44, Bolling AFB, roll #A8263, frame #0014.
13) Ibid.
14) From Headquarters, Troop Carrier Command to Commanding General, Eastern Air Command; Subject: "Report of Troop Carrier Command Participation in 'THURSDAY Operation," 03/16/44, Bolling AFB, roll #A8144, frame #0029.
15) Report, TCC participation in "THURSDAY Ops," roll #B0726, frame #0249.
16) SEAC Diaries, February 5, 1944, Memorandum from Major General Playfair to : C-in-C (11 Army Gp), Subject :LRPG Operations - 3rd Indian Division.
17) SEAC Diaries, July 11, 1944, Admiral Mountbatten to General H. H. Arnold, Subject: Combat Cargo Groups, 1,
18) Ibid., 2.
19) TCC to EAC; Subject: TCC Participation in "THURSDAY," 03/16/44, Bolling AFB, roll #8144, frame #0032.
20) History, First Air Commando Force, Bolling AFB, roll #B0680, frame #1409.
21) Report on "Op Thurs" from Air Marshal Baldwin to all commands, Bolling AFB, roll #A8143, frame #612.
22) Hdqtrs., TCC to EAC; Report of TCC during Op Thurs., 03/16/44, roll #A8144, frame #0032.
23) Bidwell's *Chindits*, 105.
24) Report on "Op Thurs" from Air Marshal Baldwin to all commands, Bolling AFB, roll #A8143, frame #612.
25) Ibid.
26) Ibid., frame #613.
27) John Masters, *The Road Past Mandalay* (New York City: Harper and Brothers; 1961), 164.
28) Air Commando l's History, frame #1410.
29) Ibid.
30) Ibid.
31) Ibid., frame #1411.
32) Report on "Op Thurs" from Air Marshal Baldwin to all commands, Bolling AFB, roll #A8143, frame #613,
33) Ibid.
34) Ibid.
35) Resume of Operation THURSDAY Bolling AFB, roll #A8144, frame #0038.
36) Ibid. Also, it must be noted that the process of "snatching" a glider from some remote airstrip gets very little play in the sources. However, in Masters' *The Road Past Mandalay* an excellent description of a glider snatch is provided. As Masters relates on page 179, "It [meaning a glider] carried with it two long poles, each

armed with a blue electric light on top-self-contained and battery operated. A third blue light shone above the pilots cockpit, Under the pilot's direction we first stuck the poles into the sand about two hundred feet in front of the glider and fifty feet apart. Then we took the glider tow rope, which was of double length, and fastened one end firmly to the tow hook in the nose of the glider. Thence we led the rope to the first pole, over the top, stretched it to the top of the second pole, and thence back to the tow hook, again fastening it firmly. The snatch plane arrived overhead flashing its recognition signal. Robbie [one of Masters' subordinates] talked to it and confirmed that we were ready, The glider pilot waited, the three blue lights burning, brakes off and controls set for take-off. The C-47 lowered under its tail a long steel boom carrying a hook with a self-operating catch. With engines throttled back, flaps full down, propellers set in full pitch, the C-47 came growling up the sandbank at about thirty feet. The hook caught in the rope, snatched it off the tops of the poles, a moment later flew taut, direct to the tow hook of the glider. The sudden jerk was taken up by the natural elasticity of the nylon, then the strain came onto the powered plane. The pilot rammed his throttles fully forward and lifted his nose. The sudden roaring growl of anguish from the engines filled the sky, The glider jerked from zero to eighty miles an hour during the stretch of the rope, perhaps a second and a half-rose in the air. The C-47 and the silent glider swept up into the darkness."

37) Ibid.

38) Air Commando l's History, frame #1411.

39) Resume of Operation THURSDAY frame #0038. Also, according to Louis Allen in his book, *Burma, the Longest War*, the Japanese Air Force commander in Burma, Major General Noboru Tazoe, on the morning of March 6th flew to Maymo to discuss his concerns regarding possible Allied attacks in the Japanese rear areas with Lt. General Renya Mutaguchi, the 15th Army commander. As noted by Allen, General Tazoe was fearful that "the Allied airborne landings could soon blossom into a tremendously strong unit in the very heart of the Japanese rear communications. They would cut off Myitkyina, and once that was taken, nothing would prevent Stilwell's road going through to China, which meant enormous pressure on the Japanese armies in China and quite possibly direct raids by US China—based planes on Japan herself. Given these implications, Mutaguchi should hurl against them every unit he could, even if it meant postponing the Imphal operation." Although Tazoe's arguments were both forceful and backed by a clear assessment of Allied capabilities, General Mutaguchi in a display of arrogance and ego minimized Tazoe's concerns and emphasized that even if a major airborne force was placed so far from their lines of supply and communications there was no way they could be possibly maintained for any length of time. Therefore, Mutaguchi dismissed Tazoe and proceeded to carry out his attacks against Imphal and Kohima. A further explanation can be found in Allen's book on pp. 326-328.

40) Report, TCC participation in "THURSDAY Ops," roll #B0726, frame #0251.

41) Air Commando l's History, frame #1412.

42) Interview, Lt. Colonel Philip Cochran conducted by Dr. J. C. Oasdorff, at Bolling AFB, 20, 21 Oct. 1975, and 11 Nov. 1975 at Rochester, NY, p.241.
43) Ibid., p.248.
44) Diary, Commanding General, Brigadier General William D. Old, Troop Carrier Command, December,1943—June,1944, Bolling AFB, roll #A8263, frame #0123.
45) Old's Diary, frame #0127.
46) Air Commando l's History, frame #1413.
47) Ibid. The United States Armed Forces designated Japanese aircraft based on the primary functions of the particular planes. Fighters, given the macho image that they represented, were assigned male names such as "Zeke" and "Oscar," whereas the more passive aircraft such as bombers and reconnaissance planes were given designations of "Betty," "Judy" and "Kate" for example.
48) Ibid.
49) Report, TCC participation in "THURSDAY Ops," roll #B0726, frame #0252.
50) Ibid.
51) Report on "Op Thurs" from Air Marshal Baldwin to all commands, Bolling AFB, roll #A8143, frame #0614.
52) Air Commando l's History, frame #1413.
53) Resume of Operation THURSDAY Bolling AFB, roll #A8144, frame #0039.
54) Old's Diary, frame #0131.
55) Ibid., frame #133.
56) Ibid.
57) Bidwell's *Chindits*,110-111.
58) Ibid.
59) Air Commando l's History, frame #1414.
60) Wingate's Report on Operations, 38-39.
61) Ibid.
62) Air Commando l's History, frame #1414.
63) Ibid.
64) Ibid., frame #1415.
65) Bidwell's *Chindits*, 193.
66) Michael Calvert, *The Chindits* (New York City, Ballantine Books,1973), 99.
67) Air Commando l's History, frame #1416.
68) Calvert's, *The Chindits*,101.
69) Air Commando l's History, frame #1416.
70) Ibid.
71) Calvert's, *The Chindits*, 105-lO6.
72) Wingate's Report on Operations, 45-46.
73) Ibid., 59-62.
74) Ibid.
75) Ibid.
76) Ibid., 64-65.
77) Ibid.
78) Ibid.

79) Air Commando l's History, frame #1417. Also, as noted in Cochran's interview, the helicopters were shipped to India in crates along with a team of engineers who would assemble them. Perhaps because of the newness of the machines and the distances travelled by sea, it would not be much of a leap to assume that time and distance might have also taken their tolls on these newfangled contraptions. Although the VR4 was used relatively late in the campaign, it is clear that with some additional upgrading of the aircraft by perhaps more durable machine parts and accounting for various climates, the potential of the machine was just beginning to be realized.
80) Ibid.
81) Ibid., frame #1418.
82) Ibid.
83) Ibid.
84) Ibid., frame #1419.
85) Bidwell's *Chindits*, 194.
86) Ibid., 154.
87) Ibid., 195.
88) Charles F. Romanus and Riley Sunderland, *United States Army in World War II-China-Burma-India Theater: Stilwell's Command Problems* (Washington, D. C.: Office of the Chief of Military History; United States Army, 1956), 195.

# CHAPTER SIX

# RESUME ON THURSDAY

However Air Commando 1 was perceived by Mountbatten and Old, within three weeks of the commencement of "Operation THURSDAY" the Commanding General of the USAAF, Henry H. Arnold, felt compelled to write another "Dear Dicky" letter to Lord Mountbatten. General Arnold had concluded that his air units in the CBI had achieved much and proved their effectiveness. Hence, General Arnold proposed to organize four more Air Commando units along the general lines of Cochran's forces and a wing of four groups especially designed "for supplying ground forces by air transportation." Included with the two proposals was a request to have both Colonel Alison and five or six of Cochran's key personnel returned to the United States to assist in the creation of these new units.[1]

.By late May, Air Commando 1 was being deactivated and sent home, as General Arnold had promised prior to its creation. Air Commando l's support of the 3rd Indian Division (originally Wingate's, but following his death, under the command of Major General W. D. A. Letaigne) came to an end on May 20, 1944. But prior to the transfer of most command personnel and turning over of support to the Northern Air Sector of the 3rd Tactical Air Force, Air Commando 1's actions for the period of February 3 until its last day, May 19, must be noted.[2]

The group's officers compiled a report on the efforts of Air Commando 1. Overall, the report indicates the enormous reliance that Wingate's units on the ground placed on Cochran's force. The P-51s (photo p.109) were called upon to fly nearly 4,500 hours of combat sorties, delivering in excess of one million pounds of bombs and incendiaries against clearly defined enemy targets. In carrying out their missions, Air Commando l's fighter pilots destroyed approximately ninety enemy aircraft on the ground and in the air, while sustaining the loss of five P-51s--one to ground fire and four lost to enemy aircraft. In addition, one B-25 was lost on a strafing mission, the only light bomber to be lost during such action.[3]

In attacks on Japanese airfields, P-51s used not only their fifty-caliber machine guns, but also wing-mounted bazookas.[4] The P-51 was proving not only its versatility, but also its adaptability in roles that its designers had probably never foreseen.

The B-25s were of course not to be outdone by the capable fighter pilots. With only ten aircraft available (it is hard to imagine what you can do with a fraction of an aircraft) at any one time as compared to twenty fighters, the B-25s were flown in combat for nearly 1300 hours, delivering an assortment of ordnance that included general purpose bombs, fragmentation cluster bombs, incendiaries, depth charges, and most significantly, 75 millimeter artillery shells fired from a mount on the nose of the bomber (a difficult weapon to use under the best of circumstances, let alone in an aircraft moving at over 300 miles per hour). The B-25s delivered a total of 431,107 pounds of all types of ordinance on the enemy. [5]

The contributions of Air Commando l's transport force were in no way overshadowed by the more "glamorous" exploits of the fighters and medium bombers. From March 5 to April 5, the transports put in some 1750 hours of flight time, using fourteen C-47s that flew 576 round trips delivering 3,456,000 pounds of freight, and flew thirteen combat sorties of double-towed gliders, as well as sixty-seven sorties of single-towed gliders. There was no loss of aircraft, but on the night of April 15th, a C-47 making a landing at "Aberdeen" managed to collide with a wayward water buffalo and was put out of action.[6]

The other units of Air Commando l's transport force, including ten UC-64s (also designated as simply C-64s), 175 gliders, and 109 Liaison aircraft, also contributed substantially to the offensive. By May, Air Commando 1 had flown fourteen glider missions with the regrettable distinction of a nearly ninety per cent attrition rate (90% of the gliders were lost in combat or in non-combat related incidents). In all, ninety-nine gliders had been released over enemy territory of which eighty-five made successful landings and fourteen crashed or were forced down. Of the seventy-five officers and twenty-five enlisted men that comprised the glider crews, ten officers and four enlisted personnel were killed. This was the highest loss rate of aircraft in Air Commando 1.[7]

The UC-64s were flown on 510 missions delivering over 800,000 pounds of combat freight with the loss of seven aircraft, but fortunately no loss of life. The light plane force of L-ls and L-5s performed spectacularly, with estimates of between 5,000 to 6,000 sorties flown evacuating more than 2,000 casualties, including seven hundred casualties brought out of the Arakan prior to "Operation THURSDAY."[8]

But the Liaison aircraft were not only removing wounded men. Some of the other functions of the force included: 1) a number of replacements for the columns; 2) reconnaissance; 3) dropping of supplies to columns, including emergency drops of medicine and ammunition; 4) evacuation of prisoners of war; 5) spotting of targets for bombing; 6) transport of mail; 7) return of captured documents and materiel; 8) transport of commanding officers and of orders. During this hectic period, forty light planes were lost, costing the lives of five pilots, of which two were caused by a mid-air collision. The vast majority of aircraft were lost either during takeoff or in landing. This loss can be attributed to the use of many airfields that were makeshift or hardly worthy of the name. They were not only small and bumpy, but had many obstacles hidden from view that caused damage to the planes. Overall, it was determined that the L-1 "proved far more suitable to an operation of this sort than the faster L-5. The L-1 on occasion carried 4 passengers, 3 stretcher cases and 1 sitting case; the L-5 sometimes carried 2 sitting cases - twice a normal load for both planes."[9]

Air Commando 1 with its small but varied force accomplished a great deal in a very brief time. However, it would be a great oversight to omit one other segment of the force. The three YR-4s that had been shipped (4 had been originally designated for shipment) to India in crates were not assembled and put into operation until the majority of "Operation THURSDAY" was mostly completed. Nevertheless, in late April, one of the helicopters was called upon to perform a rescue operation for the pilot and three passengers of a light aircraft that had been forced down due to engine failure far behind enemy lines. This was necessitated due to a total absence of any clear ground in the vicinity. The men of the downed aircraft, two of whom were injured, were directed by dropped notes to proceed to a specific ridge where they would receive supplies of food and water dropped by light aircraft.

As the downed men made their way to the appointed ridge, a YR-4 piloted by 1st Lieutenant Carter Harmen took off from Lalaghat. Due to the extremely short range of the aircraft, Lt. Harmen was compelled to make five stops before he reached Ledo, where the helicopter was fitted with an extra fuel tank so as to make it all the way to "Aberdeen." Having made it across a 6,000-foot-high ridge before landing at Ledo, Lt. Harmen was now ready for his first mission.

The stranded men were told by dropped note to come off the ridge and make their way to a small rice field nearby. Harmen's helicopter by this time had been flown to a light plane strip some five miles from where the four men were. Still in communication with a light aircraft, the men indicated that they had reached the rendezvous point and were ready for rescue. Lt. Harmen then made two flights to the clearing, bringing out the two injured men. They were then transferred to a Liaison aircraft, flown out to "Aberdeen," and subsequently flown back to India. Due to the enormous strain on the helicopter's engine from the excessive heat, the aircraft was inoperable and had to be kept at the strip overnight until its engine cooled off. The following day, Harmen made two trips to the field and brought back the other two men with no further mishaps.

This new method of rescuing men behind enemy lines was repeated. Lt. Harmen returned to "Aberdeen" on April 24th and flew four more "jobs" up to May 4th, including the pick-up of two more casualties. In total, twenty-three combat sorties were flown out of Ledo, hampered only by the tendency of the YR-4's engine to overheat. Lt. Harmen was ordered on May 4th to leave "Aberdeen" for good following a Japanese air attack that destroyed all the aircraft on the field save for the YR-4. In five harrowing days of mechanical failure, numerous thunderstorms in the mountains on one leg of the trip, and forced landings in small clearings, Lt. Harmen reached Lalaghat on the 9th of May. Having witnessed firsthand the operation of the helicopter in Burma, Colonel Cochran was led to remark that, "It really works. It's not just a stunt."[10]

Overall, Air Commando l's experiences from February through May of 1944 acted as blueprints for future operations that combined the versatility of the group's aircraft and pilots and the inventiveness of a commander like Lt. Colonel Philip Cochran. In particular, a report on the final operations of the Air Commando force went to great lengths in noting light plane operations and their future potential. Not only did the Liaison aircraft prove its usefulness in the CBI, but its operations also "foreshadowed a much greater use of such planes in this theater. By flying at tree-top level, the light planes have shown themselves almost immune to ground fire," with the requisite need for overall air superiority to ensure their safety and "The development of an improved helicopter may go far to facilitate an operation of this sort."[11] It would not be an exaggeration at this point to state that Air Commando 1 not only created a new method of warfare, but that with further refinements, this method could open up a new realm of ground-to-air cooperation that would give infantry units or airborne units true mobility.

As spectacular as Air Commando l's contributions were to the success of "Operation THURSDAY" it was not the only player on the American side to provide for the needs of the ground troops as they advanced through some of the most difficult terrain in any theater during World War II. Since Air Commando 1 was a relatively small force, it could not possibly tend to the needs of all the Allied forces in operations during the last months before the monsoon season. The Twenty-Seventh Troop Carrier Squadron in particular played an important role in "Operation

THURSDAY" and subsequent operations. During that operation, the 27th Squadron carried "412 tons of food, ammunition, and supplies, 800 soldiers and 145 mules and horses. Eighty per cent of all missions scheduled for this Squadron were completed." It had also managed to piece together two heavily damaged C-47s at CHOWRINGHEE that were about to be destroyed due to the imminent abandonment of the field, and make them serviceable enough to fly out on their own power some thirty minutes prior to the arrival of the Japanese.[12]

Once the Allies had abandoned their fields and set up their blocks at "White City," "Aberdeen," and "Blackpool" in late March and April, the most important function of TCC was to provide for the needs of the ground forces so as to maintain their long-term effectiveness in the field. The air drops were made under the cover of darkness to lessen the possibility of enemy interception and getting any advance ideas about Allied intentions. On these missions, the C-47s "carried an eight man crew consisting of the pilot, co-pilot, navigator, crew chief, radio operator and three men in the back of the airplane whose job it was to push the packs out on a signal from the pilot."[13]

To ensure that the drops hit their intended targets, Air Commando 1 drew up a set of instructions that were aimed at minimizing confusion between the cockpit and the "kickers" in the cargo bay. Ten minutes prior to reaching the drop zone, the pilot was to give warning to the ejection crew that he was coming up on the target. Approximately ten seconds before reaching the drop zone, the pilot was to switch on the red light located in the cargo bay. Upon reaching the designated drop zone, the pilot simultaneously was to turn off the red light and switch on the green light and a bell. This warning enabled the kickers to know without any verbal confirmation that they had precious few seconds to kick their cargo out the door (see photos below). This procedure of lights and bells was repeated continually as the pilot flew his aircraft in a continuous circular pattern until all of the cargo was out. Once the green light was off, all ejections were halted and the aircraft headed back home.[14]

The 27th TCS, in particular, from the moment it arrived in India in late January until the end of April, accomplished a great deal in providing the means for units on the ground to maintain their effectiveness, regardless of how far or how long they had been out of contact with their bases in India. During this period, the 27$^{th}$ accrued nearly 3,000 hours of combat flight time, which encompassed dropping 2,594,113 pounds of supplies on 567 missions out of a total of a possible 628 planned missions. In addition, the 27th brought into Burma 1,728 troops, eight jeeps, two bulldozers and one complete aircraft engine; on their return trips, the 27th flew out 427 casualties.[15] The 27th, alongside their brethren in Air Commando 1, played a crucial role in ensuring the success of the Chindits

As operations behind enemy lines were drawing to a close, it was becoming clear that an entirely new method of logistics had been attempted. The experiences of Troop Carrier Command and Air Commando 1 showed that their methods and tactics had gone beyond the experimental stage to become a proven means of maintaining large numbers of combat troops in the field for weeks. General Old in a letter of commendation to his 1st Troop Carrier Squadron noted that it had been part of an entirely new facet of air warfare. Old believed that "Air supply is revolutionizing the science of logistics, upon which all tactical successes depend. You are leaders in this revolution and you may well be proud of your part which makes ground victories possible."[16] This commendation was made three days prior to "Operation THURSDAY" so it is quite possible that TCC's commanding general was in for a surprise, as operations of the next two months would make General Old's words seem an understatement.

As a measure of the substantial impact that one of General Old's squadrons had on "Operation THURSDAY" the operational statistics of the First Troop Carrier Squadron for the period of April 5 to April 12 inclusive are significant. Within that eight-day period, the 1st TCS completed some 280 supply sorties that delivered 1,574,288 pounds of supplies and over 7200 troops with 6220 pounds of equipment.[17] Taking into consideration that the average weight of a fully loaded soldier was 185 pounds and that thirty-six soldiers were ferried on each sortie, it is truly amazing that so many men were delivered to their designated area with so few mishaps and casualties.[18]

It must also be noted that not only did the C-47s and gliders bring in combat troops, but they also on numerous occasions returned to their bases in India with cargoes of vital importance. The Liaison aircraft were instrumental in ferrying small numbers of wounded and ailing personnel out of the combat zones to waiting cargo aircraft and on occasion, gliders, for rides back to fully staffed hospital facilities in India. As Air Marshal Baldwin commented: "The air evacuation of casualties from forward areas is a service of considerable military value, since it frees fighting formations for action, and leaves them unfettered by the need to care for casualties and, moreover, it aids the recovery of casualties and reduces the time before they can return to duty."[19]

The Air Marshal's comments emphasized the difficulties that troop columns endured in caring for sick and wounded comrades while attempting to achieve their objectives. There is no question that a unit's progress in the jungle was severely hampered by having to drag along its incapacitated comrades. But Baldwin's comments have more than a ring of heartless truth to them. Units on the move rely upon speed to achieve success. Carrying along men who were unable to participate in a productive manner could put the unit's mission at risk. A cry by a seriously wounded soldier could put a unit's very existence in jeopardy. The speedy evacuation of casualties not only removed such an impediment, but also enhanced the survivability of the casualties themselves by getting them rapidly to proper medical attention, which in many cases saved their lives.

Air Marshal Baldwin went on to emphasize the pointlessness of allocating aircraft primarily for medical evacuation. He saw no reason not to use returning aircraft for evacuation purposes. And although the planes on their returning leg were far from clean and hygienic, to ensure speed and accessibility, all transport aircraft were required to carry the necessary stretcher fittings that could be folded away when not in use. He also stressed that based on inspections of the planes upon completion of their deliveries, only aircraft that had not been "carrying offensive loads" would be chosen to bring back to India its precious human cargoes. Last, medical officers on the scene were made responsible for deciding whether or not an aircraft was in such "an unhygenic [sic] condition that there [was] a danger of the wounds of the casualties being contaminated, and should there be such a danger, the decision as to whether the casualties [were] to be evacuated in that particular aircraft or wait for a possible later opportunity rests with them."[20] Air Marshal Baldwin then devised a simple method of ensuring the maximum efficiency of his planes, causing the least amount of inconvenience and at the same time providing timely transport for men in need of immediate medical care.

Overall, the accomplishments attained by the USAAF units in the CBI during the early months of 1944 were remarkable. As "Operation THURSDAY" was progressing during the months of March and April, the Japanese offensive against the major supply depot and crossroads base of Imphal, India, had resulted in fierce fighting. The Japanese launched their effort against Imphal, and Kohima, eighty miles north, in hopes of disrupting the Allies' plans and also forcing them to meet this threat to their internal lines of supply and communications. By the end of March, the Japanese had managed to tighten the noose around the defenders of Imphal. Not only were the ground forces unable to break this encirclement, but they now were forced to rely completely on aircraft to sustain them. Some 43,000 noncombatants were evacuated by air, as were 13,000 casualties. Over one million gallons of fuel, 7,000 tons of rations, over a thousand bags of mail, and more than forty million cigarettes were brought in by air.[21]

The Japanese continued to keep the pressure on the Imphal defenders throughout the following month of April. As late as the third week in April, TCC was providing the means for the men at Kohima and Imphal to be fully supplied with food, medical supplies, fresh troops, and ammunition to keep the Japanese from overwhelming the garrison. At the same time, they kept the enemy from being used against the Allied offensive taking place in north and central Burma and against the Chindit units at "Aberdeen," "White City" and "Blackpool."[22] In fact, on the three consecutive days of April 19-21, TCC flew a total of 489 sorties that delivered over 1400 tons of supplies and brought out 1875 casualties for treatment in India.[23]

But the situation at Imphal had not been resolved as the campaign dragged into May and June. Because of the enormous resources needed for the Allied offensive in Burma, it was determined that a combat cargo group would be activated on the spot to carry the burden of keeping the forces at Imphal properly provisioned and at the same time release the "borrowed" units to their previous assignments and commitments. By June 3, 1944, the first contingent of this newly activated unit, the 3rd Combat Cargo Group (CCG) arrived in India and was immediately divided into squadrons, two of which almost immediately began flying missions into Imphal. Having been officially activated on June 5th, the 3rd CCG was operating as a cohesive unit within five days.[24]

The withdrawal of Japanese forces from in and around Imphal did not end 3rd CCG's assistance to the Allied units still present in the town. Because the monsoons had arrived with their usual ferocity, ground transport in and out of Assam had come to a halt. All the roads had become impassable until early autumn. Therefore, the only way to keep the garrison healthy was to fly in all their needs. The 3rd CCG was called upon to do the job, seriously hampered by the fact that their four airdrome squadrons had not been organized as of yet (and would not be until September), requiring the crew chiefs of the aircraft to perform all the maintenance of their respective aircraft without assistance from the usual ground crews.[25]

During those first difficult months and in particular in June, the 3rd CCG had its 9th and 10th Squadrons based at Sylhet, its 11th Squadron split into two units that operated out of Dinjan and Sookerating, and its 12th Squadron operating out of Fenny in the Arakan in the south. Nearly 85% of the freight delivered by the 9th and 10th Squadrons was earmarked for Allied units at Imphal, whereas the 11th Squadron's responsibilities were to provide for Stilwell's NCAC as it made its way towards Mytikyina, and the 12th Squadron's primary concern in June was to assist the units in the Arakan as they made their advance down the coast toward Rangoon.[26]

Although the respective units would be moved around as the Allied offensive advanced during the rest of the year and into 1945, the 3rd CCG played a significant part in enabling the Allied units to reach their goals. For the ten month period of June 1944 through March 1945, the 3rd CCG amassed an enormous amount of flying time in delivering a total tonnage of supplies that equaled or exceeded the combined totals of three other transport groups, one troop carrier group, another CCG, and an RAF wing. In total, the 3rd CCG delivered nearly 160,000 tons of supplies along with almost 170,000 troops and personnel to forward areas. They also flew out more than 10,000 casualties.[27] All that was accomplished by a CCG that never existed as a unit until just before it was thrown into the fray. It may, therefore, be concluded that the United States Army Air Force units that were created and expanded for the express purpose of providing support for Allied units deployed in the enormous CBI theater not only acquitted themselves admirably, but more importantly laid the groundwork for a whole new type of warfare.

By late May, a number of dispatches sent by ground units to TCC and to the 27th TCS applauded their efforts. Brigadier General Bernard Fergusson took the time during a break in the action to pen a letter of gratitude to the 27th. While sitting on the bank of the Chindwin River as he watched his men consume American rations, General Fergusson wrote a congratulatory letter referring to his "crossing the river in boats brought by American gliders, with American outboard motors." His men were indeed "grateful and [felt] very closely akin to you all" for the efforts made on their behalf.[28]

Such feelings of gratitude by the units on the ground were expressed in voluminous detail by the SEAC commander and numerous ranking officers throughout the theater. In particular, Mountbatten was deeply grateful to the USAAF for providing the means for the units at Imphal and Kohima to maintain their tenacious defense against the Japanese offensive. As Mountbatten put it, "Although many of you were cut off and encircled, dependent on supplies dropped from aeroplanes, everyone stood firm, inspired and strengthened by the knowledge that powerful support was at hand."[29]

The commander of the Eastern Air Command, General Stratemeyer, expressed his belief that "the battle was a combined operation. If the army had not been so splendidly supported by the squadrons of the Eastern Air Command, it could not have achieved this success. Those squadrons maintained such air superiority over the battle area that hundreds of tons of all supplies could be dropped, enemy positions attacked from the air and many wounded evacuated."[30]

Indeed, the efforts of the USAAF personnel provided the means for the mainly British and Commonwealth forces to achieve their successes against the Japanese in 1944. Without Air Commando 1 and Troop Carrier Command providing assistance from the air, there is no question that the story of the CBI would have been entirely different. Starting from practically nothing and placed in an alien environment, American personnel did the impossible in just a few short months. Lord Mountbatten was pleased to convey his sincere thanks and admiration on behalf of Wingate's Chindits, and to TCC in particular, for "the smoothness, dash and efficiency with which this hazardous operation has been carried out with 100 percent success to date. All ranks of Special Force have been filled with confidence by the performance of 'Troop Carrier pilots.'"[31]

As the monsoons made their way across the subcontinent in late May and early June of 1944, the Allied advances into Burma along a 700 mile front continued unabated. Stilwell's NCAC was advancing toward Myitkyina in northern Burma, and the XIV Army under General Slim was making its way across central Burma toward their prime objective, Mandalay. The last blocking force, Major John Masters' 111 Brigade at "Blackpool," had been ordered by Wingate's successor, Brigadier Lentaigne, to abandon its positions due to enormous casualties.[32] In fact, 111 Brigade by the end of June had been reduced to approximately the size of a company, having started its mission back in March with some 3,000 effectives. As noted by Masters, "The American pilots of the Light Plane Force came, hour after hour, day after day, to the little patch of swamp we had made into a strip, and shuttled back through the heaving skies to Shaduzup."[33] Once again, the quick evacuation of stricken soldiers to medical facilities undoubtedly made the difference between life and death for many.

By the end of June, the siege of Imphal had been lifted and control of the Mogaung Valley had passed into Allied hands. At the same time, Stilwell's NCAC had captured the vital airfield at Myitkyina. It would then seem appropriate to close the book on Air Commando 1 and TCC and all the other American units that had been so instrumental in bringing about the Allied successes in Burma. The ground offensive was proceeding and Air Commando 1 had been disbanded. However, there was still a need for the use of TCC and subsequent Air Commando forces. These air units were available to provide the means to assist troops in the field with quick responses that enhanced the survivability of the wounded and also direct air support to dislodge stubborn pockets of Japanese resistance along the front.

A SEAC analysis examined the events of the previous six months and determined the successes and failures of this brand new aspect of war that today we take for granted and know as Air Mobile warfare. The report went to great lengths in identifying not only the enormous difficulties involved in this new type of warfare, but also its plusses and minuses in detail. The staff made it abundantly clear that various aspects of battle that were taken for granted in other theaters of the war were just not the same in the CBI.

For instance, because of the small sizes of dropping zones amid the terrain of Burma, compared to those available in northwest Europe, TCC was prevented from using the method known as "stick dropping" that had been standard operating procedure for the Allies. The use of "a roller conveyor running to the doorway and over lengthy dropping areas has little chance of employment."[34] Forced to deal with tiny dropping zones, it was necessary for the Allied airmen to not only be much more accurate, but also to exert themselves prodigiously over the course of a drop since more manual labor in moving crates and parcels in the cargo bay and ultimately kicking or pushing the items out was required. The dropping of fourteen tons of supplies over a zone in the Arakan in February exemplifies how proficient the airmen had become. The drop zone measured a mere fifty feet wide and only one hundred feet long. Of all the supplies that were dropped on the zone, only one package fell outside the target area.[35] Such precision was unheard of, yet the Allied airmen had honed their skills and delivered the goods.

Of course, such accomplishments did not enable the airmen to rest on their laurels. The fighting in Burma was by no means over and the need for air supply and support in the ensuing months was in no way significantly diminished. The 3rd Combat Cargo Group was activated on June 5th with the expressed assignment of "air supply for the India-Burma campaign until Mandalay could be taken and the Burma Road [Stilwell] reopened."[36] As mentioned before, the 3rd CCG was created on the spot in India following the relief of Imphal, but the necessary personnel needed to make up the four squadrons of the new group were primarily drawn from units training back in the United States, comprising both veterans and young men who had just received their wings. The 3rd CCG was officially activated on June 5th, having received its first contingent of personnel only on June 3rd, and within five days the group began operations with one squadron committed to the Hukawng Valley campaign in the north and another squadron continued the relief missions to Imphal.[37] To add to the group's difficulties, only after the middle of September, some three months after operations began, were the four airdrome squadrons up and running.[38]

But rather than be caught with inadequate personnel resources that could jeopardize the effectiveness of the 3rd CCG, extensive training programs were put into effect so as to avoid costly mistakes as operations got under way. All newly arrived pilots were immediately assigned as co-pilots, regardless of rank or prior experience. This deployment gave the men the opportunity to obtain a "feel" for flying in the mountains and over impenetrable jungle terrain. Any pilots who arrived classified as first pilots also were required to be brought up to snuff by studying flight procedure, instruments, and engines.[39] Hence, even though 3rd CCG was thrown together in a rush, it was not going to be sacrificed for the sake of speed and time.

By September, the 3rd CCG's airdrome squadrons had arrived, giving a great deal of relief to the hard-pressed air crews. Over the previous three months, the squadrons had carried out their tasks with great success, at the cost of nine aircraft lost in June, six in July, and six more in August from both accidents and enemy action.[40] Their achievements are the more remarkable since their missions throughout this period coincided with the monsoon season. In August, included in the various missions were the delivery from Chabua, India, to Myitkyina of 16-ton trailers and 20-ton tractors that comprised an entire aviation engineering battalion. Also in August, nearly three million pounds of four-inch piping and couplings were delivered to the site of the Ledo Road, which was following the advance of Stilwell's Chinese in the NCAC.[41] In addition, the 3rd CCG on its return trips to India brought out 1,875 casualties in July and 1,044 casualties in August.[42] All in all, although lacking the basic infrastructure of a fully manned combat cargo group, the 3rd managed to accomplish a great deal under less than ideal circumstances. The Official History mentions with perhaps a touch of hyperbole that "air supply has not played anywhere nearly as important a part in the European Theater as it has in the India-Burma Theater." Therefore, it was not surprising that the development of air supply methods in this theater has outstripped those other theaters.

"One division in the Burma campaign had advanced for months, fighting fiercely all the way, and depending almost entirely on air supply. And it is felt that the 3rd Combat Cargo Group, the first of its kind in operation, has justified the development of this type of air force organization."[43]

As SEAC pointed out, in many respects air transport was to some extent necessitated by the lack of mobility within the jungle environment, which at least according to their analysis, the Japanese possessed. "Allied troops do not usually possess the mobility which years of jungle warfare in China have taught the Japanese soldier; and it is air transport that provides the weapon by which the western nations might exploit their economic and technical superiority to overcome this handicap."[44]

The Allies found it necessary to dislodge a highly experienced and firmly entrenched foe in a region that could disrupt the most modest of plans. Allied command, prior to the creation of SEAC, did in fact for the first time use a form of air supply to maintain its last remaining and isolated garrison in Burma at Fort Hertz. Given that the outpost was east of the mountain range that sealed Burma off from India in the north, if the garrison was to survive, some form of supply by air was needed. On May 9, 1942, the first supply by air to this vital outpost was put into effect. Although there was no special equipment available at that time, the 2nd Troop Carrier Command made do with what was on hand, and it was able to achieve a fair amount of success. In particular, tinned foods were packed in burlap sacks of rice so as to ensure their survival once they hit the ground, and usually the dropping zones consisted of an area no bigger than that of a tennis court, usually at the "top of a conical hill."[45] This difficulty evidently gave the airmen the opportunity to acquire an accuracy heretofore unheard of in the annals of air supply.

Although this first brush with air supply was instructive and useful, the real test came when the Allies in late 1943 had begun to acquire the resources needed to implement an offensive to dislodge the Japanese from Burma. The training and research in the very specific techniques of air supply and support had initially been carried out at the Airborne Forces Research Centre in the Punjab. Experiments with various types of parachutes and materials for parachutes were carried out and tests were performed with the packing and dropping of weapons and supplies that included such items as machine guns, field howitzers, bicycles, and radio sets. "Methods of loading transport aircraft were studied, with particular attention [given] to such difficult freight as pack animals. Special emphasis was placed upon the use of materials that could be produced inside India, with an eye alike to speed in delivery and to economy in the use of articles imported from overseas."[46]

In conjunction with these trials and experiments, the RAF had been busy maintaining the Allied forces that were still in relatively remote outposts along the Indian-Burmese border throughout all of 1943. It was estimated that the units in the Chin Hills in southeast India required some

150,000 pounds of supplies each week and that the Fort Hertz position in northern Burma required in excess of 200,000 pounds of materiel every week. These nearly daily missions put a great deal of stress on the limited amount of aircraft.[47]

However, by the summer of 1943, the first C-47s had made their way to 62 Squadron. Their arrival enabled the British units to fulfill the needs of the aforementioned units and allowed them the flexibility to carry out their primary tasks, plus anything additional that might come their way. In fact, it was duly noted that the anticipated arrival of the Dakotas had been so long in coming that the members of 62 Squadron had begun to doubt that they would ever arrive.[48]

Additionally, the Air Landing School had been carrying out intense trials and training in response to the anticipated Allied buildup. As early as November of 1942, not only were they conducting live jumps of paratroopers, but perhaps more importantly they were continually modifying and refitting the containers used for various categories of supplies and also the parachutes that would be used for those containers. It was determined by January of 1943 that the Mark I, II, and III containers were ready to be put into production. Overload tests on them had proved successful, as had various modifications that would enable the troopers to handle the containers with greater ease.[49]

The Experimental Section (of SEAC) throughout 1943 continued to work on modifications of the containers and began to use Indian-made parachutes in certain circumstances. One of the trickiest problems was the handling of explosives and land mines. Through trial and error, Section C found the best and safest means for securing these highly delicate and dangerous supplies was by modifying packing materials and also putting less in a given container and filling the excess space with materials that could absorb the shock of hitting the ground via parachute. Drop tests with jellified acid batteries in the new Mark V containers were carried out successfully, enabling radio sets to be airdropped rather than flown in by a single aircraft.[50]

By late August 1943, 62 Squadron had begun to receive C-47s beyond its initial complement of four. Nine additional aircraft brought the total to thirteen planes, which put an enormous strain on the ground crews. This inconvenience was only temporary; the number of ground personnel was increased to meet the needs of the unit.[51]

Even as support for an audacious airborne assault was gathering momentum throughout the theater, skepticism was still plaguing Allied planners. In particular, the British Army in India raised doubts as to the feasibility of such an undertaking. To convince the doubters of the correctness of the proposed assault, throughout the month of December, the RAF showcased the plans for Army commanders, elucidating instructions, preparations, and above all actual demonstrations of the planned assault, "Operation THURSDAY." It was hoped that any lingering doubts would be swept away in an avalanche of efficiency and daring.[52]

Once the bulk of Troop Carrier Command had arrived along with Air Commando 1, the Allies had the means to put these techniques to use. As noted by SEAC, as early as February 1944, in the Arakan alone over ten thousand short tons of supplies had been delivered to the 5th, 7th, and 81st Divisions via approximately 3,000 sorties. In fact, during the most intense period of February 7th through the 20th, some 640 sorties were flown by transport aircraft of both the USAAF and the RAF, delivering "1,303 British tons of which 29.7% was dropped free and 70.3% parachuted. Men and rations accounted for the greater part of this weight with 805 1/2 tons and ammunition came next with 273 1/4. The despatch [sic] of these men and supplies by road from the railhead at Dohazari had the way been open would have taken approximately 434 trips by 3-ton lorries, each trip lasting for at least twelve hours."[53] Therefore, it can be said that by the time "Operation THURSDAY" got underway, many of the difficulties had been ironed out, including the unification of command and the expressed intent of carrying the offensive to the enemy, not only at the frontline, but also and more importantly, in the more vulnerable rear areas.

As previously noted, the success of "Operation THURSDAY" hinged on the ability of TCC and Air Commando 1 to supply the airborne troops while they were, in some cases, 160 miles from the Allied frontlines. And although the entire mission was carried out with little if any interference from the Japanese, it did not mean that SEAC did not encounter various problems along the way. In one particular incident, a problem resulted from "the refusal of a well-known mule, Cushy Charlie, to be emplaned. He successfully resisted the efforts of thirty-five men directed by General Wingate in person."[54] SEAC failed to mention the results of Wingate's intervention, but it can easily be surmised that Cushy Charlie did not hold up the entire mission for very long.

Overall, SEAC summed up "Operation THURSDAY" by pointing out that "a total of 9,250 men, together with eleven or twelve hundred animals, complete with weapons and equipment and full organization [had been landed]. The transport planes that achieved this result were drawn not only from the First Air Commando, but also Dakotas from the 27th and 315th Troop Carrier Squadrons of the U.S.A.A.F., and the 31, 117, and 194 Squadrons of the R.A.F."[55] The mission succeeded beyond anyone's most optimistic expectations. But this was only the beginning of a long and arduous campaign to push the Japanese out of central Burma, including the key central city of Mandalay and ultimately the capital, Rangoon. What the Allies needed at that point may not have been of the same nature as "Operation THURSDAY" but what they had learned and achieved would come in handy for future operations regardless of where the offensive would take them.

Coupled with the growing efficiency of delivering men and material to where they were needed most, the evacuation of casualties took on an importance not originally envisioned. Empty aircraft could be used on their return trips to carry wounded or sickened troops safely behind the front lines to medical facilities unavailable to units in the field. As early as April of 1942, the 31 Squadron of the RAF gave frequent help in the evacuation of casualties during the headlong retreat of British forces from Burma. On one day alone in April, 31 Squadron flew 188 casualties and 90 evacuees from Myitkyina to safety in Dinjan in the Assam region of India.[56]

As 31 Squadron continued to provide a bare-bones operation of emergency medical evacuation, other British units and subsequent American units joined in. The 117 Squadron had been instrumental in evacuating casualties from Sicily in July and August of 1943 and was shortly thereafter transferred to the CBI. And a pilot from 31 Squadron, during the first Wingate expedition, single-handedly made a daring landing in the jungle and took aboard his C-47 seventeen sick and wounded Chindits. Little by little, Allied units were gaining experience, but the major proof of their training and skills came with the launching of the Japanese invasion of India at the two key points of Imphal and Kohima in February of 1944. Because Allied units were cut off from their rear areas and had very little open space available, the use of the newly arriving liaison aircraft of the USAAF proved to be the key to success in removing casualties to superior facilities further to the rear. An entire system of ferrying the wounded from the lines of battle to the makeshift airfields where the liaison aircraft could land and take off, and then to an airfield close to the battlefields but big enough to handle C-47s was developed that enabled the Allies to evacuate 666 casualties from the Arakan in February.[57]

From their experiences in the Arakan alone, it was determined that a C-47 was capable of being fitted with sixteen fold-down stretchers that could be stowed away during supply operations, and that from eight to twelve sitting cases could also be accommodated on a flight back to a rear base. It was also determined that to effect a high survival rate for the casualties, a maximum of twenty minutes was all that was necessary to load the plane to its capacity.[58]

Therefore, coupled with the experiences of the RAF plus the arrival and subsequent intensive training of USAAF units, as well as experiences gained prior to "Operation THURSDAY" the Allies were prepared for the air assault behind Japanese lines in March of 1944. By the time the operation commenced, the evacuation of casualties had become almost routine. Between March 5th and April 4th, some 278 casualties (the relatively low number was due to the minimal amount of contact that the airborne units had with the enemy) were flown out of Burma to medical facilities in India. And as impressive a figure as that was, during the entire month of April some 8,996 casualties were flown out of all areas of combat throughout India and Burma. The following month of May saw 8,593 casualties removed from the forward areas, which reflected not only a high rate of efficiency on the part of TCC, but also the importance that SEAC placed on getting the men quickly to proper medical facilities. Overall, in the first five months of 1944, TCC flew 23,209 wounded and sick combat soldiers out of danger and directly to forward hospitals in India. In fact, because of the enormous strain that those many casualties was putting on forward hospital facilities, TCC flew several hundred casualties directly to Calcutta from Chittagong, thereby obviating congestion at hospitals closer to the front.[59]

It must also be noted that the introduction of the helicopter, though notable, was not a major factor, because of its quirky nature and vulnerability to inclement conditions, specifically its tendency to overheat, especially in hot and humid conditions. As newer models made their way into combat areas and modifications were made to make these aircraft less susceptible to breakdowns and malfunctions, they would be able to play a much more significant role.

However, one aircraft that may seem out of place in the jungles of Burma came to play a significant role, especially as the monsoons came into full swing by early June, when finding relatively hard strips or all-weather landing strips was exceedingly difficult. If the C-47 was mired in the mud, then not only would the casualties suffer, but most likely the aircraft would have to be abandoned. However, Burma was a country with numerous large and small rivers flowing through it. Therefore, SEAC enlisted the use of flying boats that had been based in Ceylon (known today as Sri Lanka) and had them flown up to waters of the upper Brahmaputra River, where they were stationed on an on-call basis. They were then flown into Burma to the Irrawaddy River and Indawgyi Lake so as to effect a reasonably quick method of air evacuation of casualties. The

3 Indian Division, operating southwest of Myitkyina, met with these flying boats on the lake and were flown back over the mountains to Assam. Otherwise, the casualties would have been evacuated overland, and their survival would have been seriously in doubt. These flying boats received an escort of Spitfires to keep enemy activity in the air and on the ground at bay.[60]

## Notes

1) Letter, General H. H. Arnold, CO-USAAF to Lord Louis Mountbatten, Commander-in-Chief, Southeast Asia Command, March 24, 1944, Henry H. Arnold Papers, Manuscript Division of the Library of Congress, Washington, D.C.
2) "Report on Final Operations of First Air Commando Group," located at Bolling AFB, Roll #A8144, frame #0005.
3) Ibid., frame #0006.
4) Ibid.
5) Ibid.
6) Ibid., frame #0007.
7) Ibid., frame #0009.
8) Ibid.
9) Ibid., frame #0010.
10) Ibid.
11) Ibid., frame #0012.
12) Ibid., frame #0013.
13) 27th TCS History, frame #0437.
14) Air Commando I's History, frame #1443.
15) Air Commando I's History, frame #1444.
16) 27th TCS History, frame #0441.
17) First Troop Carrier Squadron Office of the Intelligence Officer, commendation letter
from Commanding General, TCC, to 1st Troop Carrier Squadron, dated March 2, 1944, located at Bolling AFB, Roll #A0966, frame #1862.
18) Ibid., frame #2056.
19) Ibid., frame #2055.
20) Third Tactical Air Force, Operation Instruction No. IS/44, Air Evacuation of Casualties, Bolling AFB, Roll #A8143, p. 1.
21) Ibid., pp. 2-3.
22) Louis Allen, *Burma: The Longest War,1941-1945* (New York City: St. Martin's Press; 1984), 244.
23) Old's Diary, frame #0151.
24) "Summary of 3rd Combat Cargo Group," April, 1945, Bolling AFB, Roll #B0726, frame #0059.
25) Ibid.
26) Ibid., frame #0070.
27) Ibid., frame #0078.

28) Letter, Brigadier General Bernard Fergusson "To all the American Officers and men who sent us comforts," March 9, 1944, provided by Harry A. Blair, Historian, 27th Troop Carrier Squadron.
29) 27th TCS History, frame #0630.
30) Ibid., frame #0631.
31) Ibid., frame #0632.
32) Allen's *Longest War*, p. 379.
33) Masters' *Road Past Mandalay*, p. 259.
34) "Air Transport Operations on the Burma Front" prepared by the staff of SEAC July, 1944, SEAC War Diaries, National Archives, Suitland, Maryland, 4.
35) SEAC Diaries, Staff Analysis, p.5.
36) "3rd Combat Cargo Group Statistical Survey, June-November, 1944," Bolling Air Force Base, Washington, D.C., Roll #A0966, frame #0056.
37) Ibid., frame #0059.
38) Ibid.
39) Ibid., frame #0060.
40) Ibid., frame #0070.
41) Ibid.
42) Ibid., frame #0071.
43) Ibid., frame #0038.
44) Ibid., frame #0086
45) SEAC Diaries, p.6.
46) Ibid., p.7
47) "Air Supply Commitments" from Air Headquarters, Bengal to 0. C. 31 Squadron, Kharagpur, 21st June 1943, located at the Public Record Office, London, England; Air 23/2021, "Appendix "A."
48) "Summary of Events," Operations Record Book, No. 62 Squadron, 24/7/43, located at PRO, Air 27/583, p.138.
49) "Summary of Events," Operations Record Book, Air Landing School, June, 1943, located at PRO, Air 29/511.
50) Ibid.
51) 62 Squadron's Record Book, August 13—16, p.141.
52) "Summary of Events," Operations Record Book, No. 117 Squadron, November, 1943, located at PRO, Air 27/903, p.21.
53) SEAC Diaries, p.9.
54) Ibid., p.14.
55) Ibid., p.16.
56) Ibid., p.18.
57) Ibid., p.25.
58) Ibid.
59) Ibid., p.26.

# CHAPTER SEVEN

# JUNGLE TO JUNGLE

In the first six months of 1944, the Allies developed techniques and methods under extremely harsh conditions, laying the groundwork for an entirely new method of warfare. Taking into consideration the enormous difficulties in melding two entirely different air forces under a new command structure, the successful implementation of these techniques is all the more remarkable. Air mobility was honed to a fine art by the men in TCC and Air Commando 1.

In fact, it can be unequivocally stated that in the backwaters of World War Two, away from most of the attention of the world, American and British airmen created something out of nothing and put into existence a new form of warfare that could only be imagined a few short years before. With the appropriate tools and the desire to succeed in their mission, and at times totally ignoring conventional wisdom, SEAC's forces accomplished a great deal. Air Commando 1, a unique creation in its own right, put standard operating procedures aside and at times abandoned tradition. It operated as an extension of the Chindit units on the ground. Its close-in tactical air support of the ground forces was by no means unheard of, but its abilities to give the brigades on the ground what they needed and practically on a moment's notice ensured not only the success of the units behind the Japanese lines, but proved that units could indeed cooperate on such a level.

The dawn of a new type of warfare had been born in the jungles of Burma. There is no question that airborne operations in Sicily, Italy, France, and Holland received a great deal more play in the press and within the armed forces of the United States and Great Britain. However, aside from the gallant resistance of the 101st Airborne at Bastogne in December of 1944,[1] airborne units were not meant to be left on their own for more than a few days before they linked up with ground units.[2] The Allied units flown into the rear of the Japanese were assigned to be there for weeks on end, totally dependent for their survival on the guardian angels of Air Commando 1 and TCC.

Also, Air Commando 1 perfected such close-in tactical air support of the ground units that it was possible to call in a strafing run of P-51s and not only get a response in less than one hour, but have the planes lay down fire on targets within 150 yards of ground positions.[3] This accomplishment is even more noteworthy given the lesser progress of close-in tactical air support provided by Allied air units to ground units in France following the landings in Normandy. As noted in Michael Doubler's book, *Closing With the Enemy*, fighters maintained a minimum of 500 yards between friendly forces and the enemy. As the forward observers and pilots became more experienced and confident, fighter-bombers were able to strike within 300 yards of friendly territory.[4] In the CBI, Allied aircraft had been effectively attacking Japanese targets significantly closer to friendly lines, with little hazard to Allied troops on the ground.[5] The dawn of a new type of warfare had been born in the jungles of Burma.

The role of aircraft in World War Two proved beyond any doubt the importance of both strategic and tactical aircraft. Despite great emphasis on the American 8th Air Force flying out of bases in England and the dominant role that carrier-based aircraft power played in the Pacific, it was in the CBI that the blueprints and framework of an entirely new form of warfare were tried and proved effective. Although many of the methods, if attempted against a heavily armed enemy (such as the Germans, in particular during "Operation Market-Garden") might have resulted in mass slaughter of airborne units, it has to be emphasized that the role played by airborne units in the CBI was specifically to avoid encountering "hard targets." The Chindits were lightly armed and, given the terrain, extremely mobile. Therefore, they could only be effective by striking at the enemy where they least expected it and moving on as quickly as possible. "Operation THURSDAY" proved that the airborne units could wreak havoc behind the Japanese lines and rely on TCC and Air Commando 1 for all their needs. When their mission was over, they returned safely with the assistance of their guardian angels based in India.

There was undoubtedly a linear connection between the jungles of Burma and the jungles of Vietnam. Air cavalry units ferried by helicopter throughout the length and breadth of South Vietnam in the 1960s and early 1970s were no different in essence from the Chindits and other airborne units in the CBI. Had the Allies had at their disposal the varied helicopters and other aircraft that the United States Army, Navy, and Marines had in Vietnam, there is no telling what they might have accomplished. Air mobility was proved to be not only another facet of an all-out offensive against an entrenched enemy, but it also gave the wounded or sickened combat soldier a far better chance of survival than he could ever have hoped for.

It must also be noted that in cooperation between two Allied forces in the field, the CBI theater showed that it could be done with a minimum of rancor and suspicion from both sides. Yes, the Allied coalition under General Eisenhower succeeded in defeating the Germans in western Europe and hastened the destruction of Nazi Germany. But in reading the plethora of biographies and memoirs of the principal players, it is easy to see jealousy and suspicion just beneath the surface. That is not to say that the CBI was completely free of such antagonisms (considering the loathing that Lt. General Joseph Stilwell had for his British Allies[6] ) and jealousies. However, with the predominance of American air power and British Commonwealth and Chinese ground units that relied upon American air power, it seemed that SEAC had an advantage that was just not possible in Europe.

Overall, given the enormous difficulties in logistics, climate, and geography, it is amazing what the Allies managed to accomplish against the Japanese. Although General Stilwell's son, Colonel Joseph Stilwell, Jr., described his and his father's experience in the CBI in a report entitled, "Confusion Beyond Imagination," a great many accomplishments came out of that confusion. The CBI was the birthplace and the proving ground for an entirely new method of warfare. Air mobility was not only proved feasible, but it could also be sustained, theoretically for as long as needed, provided the resources were available.

It is regrettable that the CBI has been neglected in American history. To the British, on the other hand, Burma was the theater that comprised the greatest commitment of time and resources, considering that they and the Japanese were locked in combat over a large piece of territory for over three years. The numerous secondary source materials cited in this book are from British authors; few were written by Americans.

The little that people in the United States know of the CBI stems primarily from the events surrounding the 5307th Composite Group, also officially designated as "Galahad" and more commonly known as "Merrill's Marauders." A movie made in the 1940s starring Errol Flynn, "Destination Burma," relates some of the experiences of that unit, but regrettably implies that it was primarily United States ground forces that defeated the Japanese. However, the movie has one scene that most likely cannot be found anywhere else on celluloid. It accurately depicts the snatching of gliders from a remote airfield in the middle of the jungle.

In a war during which nearly all of the industrialized world was engulfed in horrific conflict, the CBI became a mere footnote in regard to

American involvement in World War Two. This researcher's intent is to shed some light on the CBI and bring to the fore an awareness of something so unique and special that to leave it buried in the archives is a disservice to the men who served there and their accomplishments. It also ignores a basic aspect of historical analysis.

The first steps in developing an entirely new method of warfare so heavily used another place and time, Vietnam, were taken in the jungles of Burma. That deployment was not far from a stone's throw from its origins in another jungle environment known as the CBI.

## Notes

1) On December 17th, 1944, the 101st Airborne Division was ordered to deploy by ground transport to the key road center of Bastogne. It can be stated that this was more of an ad hoc deployment of a much needed force of proven effectiveness than as a specific airborne operation. Nevertheless, the "Screaming Eagles" were compelled to hold against the German encirclement and subsequent pounding for nearly four weeks. To learn more of the 101st's determined defense of Bastogne against unrelenting pressure, both Charles B. MacDonald's, *A Time for Trumpets* (New York City: William Morrow and Company, Inc; 1985), and Danny S. Parker's *Battle of the Bulge* (Philadelphia: Combined Books; 1991), give very thorough accounts of their experiences.

2) As Cornelius Ryan points out in his book, *A Bridge Too Far* (New York City: Popular Library; 1974) , the furthest penetration of Allied airborne units was to be some sixty-three miles from the Allied lines (near the Dutch-Belgian border) at the bridge located in the town of Arnhem, the Netherlands. As it turned out, the British 1st Airborne Division never was able to link up with General Horrick's XXX Corps and ultimately was forced to withdraw suffering not only defeat, but embarrassment of having been ordered to retreat under the cover of darkness and having failed to hold onto their prize, the bridge that spanned the Rhine and access to the industrial heartland of the Third Reich.

3) "Unit History of the First Air Commando Force," Bolling Air Force Base, Washington, D. C., Roll #B0680, frame #1667-70.

4) Michael Doubler points out in his recent book, *Closing With The Enemy* (Lawrence, Kansas: University Press of Kansas; 1994), that tactical air support of Allied units in the Normandy beachhead initially was tentative and caused numerous friendly fire casualties. However, once the ground units with appropriately experienced forward observers and experienced fighter pilots in the air, casualties were minimized and tactical air became enormously effective. This points out that although 21st Army Group had at its disposal overwhelming ground and air resources to dislodge the Germans from the bocage (hedgerows) countryside, it was the time honored process of practice by doing that ultimately brought about successful coordination between air and ground forces. It can be surmised that Air Commando 1 had been more effective with less time to hone its skills than the Allied air forces in northwest Europe. I contend that Cochran's men

were much more flexible in their doctrine and were much more willing to try anything that might prove effective in keeping the Japanese at bay against the Chindits.

5) Air Commando l's History, frame #1671.

6) Throughout *The Stilwell Papers* edited by Theodore F. White (New York City: William Sloane Associates; 1948), General Stilwell vents his spleen continually on his enormous disgust for his "Limey" Allies. His vitriolic contempt toward his British cousins most certainly had to affect his judgment in some crucial circumstances, regardless of the final outcome.

# BIBLIOGRAPHY

## Primary Sources


Library of Congress, Manuscript Division, Washington, D.C.: General of the Air Force, Henry Harley Arnold papers.

National Archives, Suitland, Maryland:

SEAC war diaries including daily reports of all Allied units in the field; Future plans developed by SEAC staff, 1943—1945; Communiques between Headquarters, SEAC to British Air Ministry, London and Joint Chiefs of Staff and Air Ministry, British Military Staff, Washington, D.C.; Extracts from Supreme Commander's Lord Louis Mountbatten personal diary; Extracts from weekly intelligence summaries; Monthly summations prepared by staff of available men and material and local production goals.

National Archives and Records Administration, Washington, D.C.

United States Army Air Corps records, 1919-1943.

1) All Army Air Corps Records listed under heading, Record Group #18.
   a) File #452.3: General history of Air Corps and Air Support Staff Section,1917-1938.
      i) Box #s 1045—1048, 802, 803.
   b) File #360.01: Development of Aviation for Close Support of Ground Troops, 1941-1942.
      i) Box #s 163-166, 179-181, 134-139.
   c) File #s 322.3-322.7-Future Acquisitions of Aircraft and Proposals for New Technologies.
      i) Box #s 138-139.
   d) File #s 312.1, 313.3, 319.1, 321.9, 333.1, 344.2, 350.001, and 370-"Establishment of Policy."
      i) Box #s 1883-1892.

Historical Branch, G-3 Division, Headquarters, U. S. Marine Corps, Washington, D. C. Marine Corps Historical Reference Series, Number 21, "The United States Marines in Nicaragua."

Imperial War Museum, Department of Documents: Wingate Papers; London, England.

Public Record Office, RAF records, London, England. Specific After Action reports of Royal Air Force #31 Combat Air Wing, #62 & 194 RAF Squadrons, and 117th Brigade records.

1) Air 23/1943, First Wingate Operation.
2) Air 23/2021, Supply Dropping, Policy.
3) Air 23/2035, Operation of 31 Squadron.
4) Air 29/511, Operations and Record Book of the Air Landing School, Chaklala
5) Air 23/5386, Parachute Training School at Ringway in the United Kingdom.
6) Air 29/513, Airborne Forces Experimental Establishment Ringway, UK.
7) Air 29/515, AFEE Jan-Jul 1943.
8) Air 23/2486, "A Review of Air Transport Operations on the Burma Front to June 1944."
9) Air 23/5154, Supplies by Air Dropping Operations, 1942-43. United States Air Force Archives at Bolling Air Force Base, Washington, D.C. 10th Air Force and Air Transport Command records, including statistical surveys of 3rd Combat Cargo Group, 1943-1945:
    1) Monthly unit histories.
        a) 1st Air Commando Group
        b) 3rd Combat Cargo Group
            i) 11th Combat Cargo Squadron
            ii) 27th Troop Carrier Squadron
    2) After-action reports of 10th Air Force.
        a) Campaign summaries of 3rd Combat Cargo Group.
        b) Statistical reviews of squadrons and groups.
    3) Troop Carrier Command reports of the Office of the Intelligence Officer.
    4) Tenth Air Force Weight of Effort Quarterly Reports, 1944-1945.
    5) Air Command, SEAC staff generated report on air transport operations, July,1944.
    6) Reviews of Air Operations against specific targets, quarterly generated.
    7) Chronological History of 10th Air Force's Air Supply Operations, 1943-1945.

Taped interview of Colonel Philip G. Cochran, commander of Air Commando 1, by Dr. James C. Hasdorff, 10/20/75—11/11/75.

Diary of the Commanding General of the Troop Carrier Command, William D. Old, located on Roll #A8263.

Taped interview of Major General Howard C. Davidson, commander of the Tenth Air Force, CBI, December,1974.

United States Army Military History Institute, Carlisle Barracks, Pennsylvania.

## Interviews and Correspondence

Carre, Walter. Enlisted man in 7th Bomb Group based in India in spring of 1943 and transferred to China in late 1944. Interviewed on May 21-22, 1993 in King of Prussia, PA.

Flint, Ralph J. Enlisted man in 3rd Combat Cargo Group based in India from spring of 1944 and was transferred to China with unit in early 1945. Interviewed on May 21-22, 1993 in King of Prussia, PA.

Quinn, William S. Enlisted man in the 1st Air Cargo Resupply based in India from spring of 1944 through early 1945. Interviewed on May 21-22, 1993 in King of Prussia, PA.

Blair, Harry A. Enlisted man of the 27th Combat Cargo Squadron, unit's official historian who provided me with his own experiences and various records pertaining to the 27th upon their arrival in India in early 1944. Our exchange of letters and phone calls began in December of 1992 and continued until April of 1993. Woodruff, Woodrow B. Jr. Enlisted man served with the 612 Field Artillery Battalion as part of the Mars Task Force following the dissolution of Merrill's Marauders in the fall of 1944. Mr. Woodruff responded to a mailing of over 200 questionnaires sent out to CBI veterans that was provided to me through the CBI Veterans of America Association headquartered in Parma Heights, Ohio. Mr. Woodruff was extremely generous in providing an additional list of names of veterans who he believed would provide me with additional information. Our exchange of letters began in November of 1992 and concluded by March of 1993 in which he included a number of photographs sent to him by the widow of Orfeo Bonardi, who trained with him at Fort Sill during the winter of 1943-44. Mr. Bonardi, serving with Battery A of the 612 Field Artillery received a head wound that disabled him permanently and he died in 1991.

## Secondary Sources

Allen, Louis. *Burma: The Longest War.* New York: St. Martin's Press, 1984.

Ambrose, Stephen E. *Citizen Soldiers.* New York: Simon and Schuster, 1997.

—. *D-Day*: *June 6, 1944*: *The Climactic Battle of World War Two* New York: Simon and Schuster, 1995.

Angelucci, Enzo. *The Rand McNally Encyclopedia of Military Aircraft.* New York: The Military Press, 1980.

Barker, A. J. *The March on Delhi.* London: Faber & Faber, 1963.

Bidwell, Shelford. *The Chindit War.* New York: Macmillan Publishing, 1979.

Booth, Michael T. *Paratrooper.* New York: Simon & Schuster, 1994.

Callahan, Raymond. *Burma 1942-1945.* New York: Davis-Poynter, 1978.

Calvert, Michael. *The Chindits.* London: Ballantine, 1973.

—. *Fighting Mad.* London: Jarrolds, 1964.

Calvocoressi, Peter and Winty, Guy. *Total War.* Middlesex, England: Penguin Books, 1972.

Chappell, John. *Before the Bomb: How the United States Approached the War in the Pacific.* Lexington, Kentucky: University Press of Kentucky, 1997.

Crablehouse, Charles. *Helicopters and Autogiros.* Philadelphia: J. B. Lippincott Co., 1967.

Craven, Wesley Frank and Cate, James Lea. *The Army Air Force in World War Two, volume IV.* Chicago: University of Chicago Press, 1953.

—. *The Army Air Force in World War Two.* Chicago: University of Chicago Press, 1955.

Doubler, Michael D. *Closing With the Enemy: How GIs Fought the War in Europe, 1944-45.* Lawrence, Kansas: University of Kansas Press, 1994.

Ehrman, John. *History of the Second World War, Grand Strategy, volume V.* London: Her Majesty's Stationery Office, 1956.

—. *History of the Second World War, Grand Strategy, volume VI* . London: Her Majesty's Stationery Office, 1956.

Eldridge, Fred. *Wrath in Burma.* Garden City, New York: Doubleday, 1946.

Fay, John. *The Helicopter: History, Piloting & How It Flies.* New York: David & Charles, 1987.

—. *The Helicopter: History, Piloting and How It Flies.* Vancouver: David & Charles, 1954.

Ferret, Geoffrey. *Winged Victory: The Army Air Forces in World War II.* New York: Random House, 1993.

Howard, Michael. *History of the Second World War, Grand Strategy, volume IV.* London: Her Majesty's Stationery Office, 1970.

Hunter, Charles N. *Galahad.* San Antonio, Texas: The Naylor Company, 1963.

Kirby, S. Woodburn. *The War Against Japan: India's Most Dangerous Hour, volume II:* London, Her Majesty's Stationery Office, 1958.

Mabaulay, Neil. *The Sandino Affair.* Chicago, Illinois: Quadrangle Books, 1967.

Masters, John. *The Road Past Mandalay.* New York: Harper and Brothers, 1961.

Mitchell, William. *Skyways.* Philadelphia: J. B. Lippincott, 1930.

—. *Winged Defense.* London: G. P. Putnam's Sons, 1925.

Munson, Kenneth. *Helicopters and Other Rotorcraft Since 1907.* London: Blandford Press, 1973.

O'brien, Terence. O*ut of the Blue.* London: Collins, 1984.

Ogburn, Charlton. *The Marauders.* New York: Harper and Brothers, 1959.

Ohl, John Kennedy. *Supplying the Troops: General Somervell and American Logistics in World War I.* Dekalb, Illinois: Northern Illinois University Press, 1994.

Prasad, Bisheshwar, ed. *The Official History of the Indian Armed Forces in the Second World War, volume 1.* Calcutta: Sree Saraswaty Press, 1958.

—. *The Official History of the Indian Armed Forces in the Second World War, volume* 2. Calcutta: Sree Saraswaty Press, 1959.

Romanus, Charles and Sunderland, Riley. *The U.S. Army in World War II, The China-Burma-India Theater: Stilwell's Mission to China.* Washington, D. C.: Department of the Army, 1953.

—. *The U.S. Army in World War II, The China—Burma—India Theater: Stilwell's Command Problems.* Washington, D. C.: Department of the Army, 1956.

—. *The U.S. Army in World War II, The China-Burma-India Theater: Time Runs Out in CBI.* Washington, D. C.: Department of the Army, 1959.

Slim, Field Marshall the Viscount William. *Defeat Into Victory.* New York: David McKay Company, 1961.

Sykes, Christopher. *Orde Wingate* London: Collins, 1959.

Taylor, Dr. Joe G. *Air Supply in the Burma Campaigns.* Washington: USAF Historical Division, 1981.

Thomas, Lowell. *Back to Mandalay.* New York: Greystone Press, 1951.

Tuchman, Barbara. *Stilwell and the American Experience in China: 1941-1945.* New York: Macmillan Company, 1971.

Tulloch, Major General Derek. *Wingate in Peace and War.* London: Macdonald and Company, Ltd., 1972.

White, Theodore H. ed. *The Stilwell Papers.* New York: William Sloane Associates, 1948.

# INDEX